Factories of the Future

Scrivener Publishing
100 Cummings Center, Suite 541J
Beverly, MA 01915-6106

Publishers at Scrivener
Martin Scrivener (martin@scrivenerpublishing.com)
Phillip Carmical (pcarmical@scrivenerpublishing.com)

Factories of the Future

Technological Advancements in the Manufacturing Industry

Edited by

Chandan Deep Singh

and

Harleen Kaur

Scrivener
Publishing

WILEY

This edition first published 2023 by John Wiley & Sons, Inc., 111 River Street, Hoboken, NJ 07030, USA and Scrivener Publishing LLC, 100 Cummings Center, Suite 541J, Beverly, MA 01915, USA
© 2023 Scrivener Publishing LLC
For more information about Scrivener publications please visit www.scrivenerpublishing.com.

Wiley Global Headquarters
111 River Street, Hoboken, NJ 07030, USA

For details of our global editorial offices, customer services, and more information about Wiley products visit us at www.wiley.com.

Limit of Liability/Disclaimer of Warranty
While the publisher and authors have used their best efforts in preparing this work, they make no representations or warranties with respect to the accuracy or completeness of the contents of this work and specifically disclaim all warranties, including without limitation any implied warranties of merchantability or fitness for a particular purpose. No warranty may be created or extended by sales representatives, written sales materials, or promotional statements for this work. The fact that an organization, website, or product is referred to in this work as a citation and/or potential source of further information does not mean that the publisher and authors endorse the information or services the organization, website, or product may provide or recommendations it may make. This work is sold with the understanding that the publisher is not engaged in rendering professional services. The advice and strategies contained herein may not be suitable for your situation. You should consult with a specialist where appropriate. Neither the publisher nor authors shall be liable for any loss of profit or any other commercial damages, including but not limited to special, incidental, consequential, or other damages. Further, readers should be aware that websites listed in this work may have changed or disappeared between when this work was written and when it is read.

Library of Congress Cataloging-in-Publication Data

ISBN 978-1-119-86494-3

Cover image: Pixabay.Com
Cover design by Russell Richardson

Set in size of 11pt and Minion Pro by Manila Typesetting Company, Makati, Philippines

Printed in the USA

10 9 8 7 6 5 4 3 2 1

Contents

Preface

As the title suggests, this book covers the factories of the future and other technological advancements in the manufacturing industry. Basically, the book is divided into two parts: emerging technologies and advancements in existing technologies. The chapters on emerging technologies consist of topics on Industry 5.0, machine learning, intelligent machining, advanced maintenance and reliability; whereas the chapters on advancements in existing technologies cover digital manufacturing, artificial intelligence in machine learning, the internet of things, product life cycle, and the impact of factories of the future on the performance of manufacturing industries. Since these technical advancements are not currently available in a single book, an attempt has been made to cover all these topics in one book.

Presently, a major concern of all manufacturing industries is to remain agile in order to be competitive in the market. To achieve this goal, industries have to adopt new technologies in order to provide customers with better quality products and reduce lead time. In addition to this, a subsequent issue facing manufacturing is helping to save the environment with the introduction of "green" practices, otherwise known as "green manufacturing."

Nowadays, even research is concentrating on these technical advancements; thus, the intended audience of this book will find material regarding these advancements collected in a single volume. Furthermore, they will develop an understanding of the social, economic and technical justifications for these advancements. In addition to this, the role different technologies play with each other is investigated.

This book will be useful to researchers, academics and faculty in industrial production, mechanical engineering, electronics and other allied branches of manufacturing, and even those in business analytics programs, as case studies and analytical tools related to manufacturing performance are provided. It will also be helpful to those in industrial R&D departments, as industries are always adopting new technologies as advancements are continually being made in this sector. So, on the whole, it will be helpful to both academicians and industrialists.

Technological advancements in manufacturing is a pressing topic, as industries have to sustain their performance when adopting advanced production techniques, and while doing so, they have to keep green issues in mind. Therefore, this is a vastly important book as it involves improving the performance of businesses while creating technical advancements, thus making them more competitive. Since this book contains detailed information on the technological advancements made in manufacturing production and maintenance, it will provide insights into the various technologies currently adopted by industries and those yet to be adopted, that will impact the future performance of industrial manufacturing.

Dr. Chandan Deep Singh and Dr. Harleen Kaur
December 2022

Factories of the Future

Talwinder Singh* and Davinder Singh

Department of Mechanical Engineering, Punjabi University, Patiala, Punjab, India

Abstract

Rapid progress of smart technologies result in drastic changes in industrial production processes thereby building a roadmap for "The Factories of the Future" with a clear vision of how producers should improve productivity through advancement in plant structure, plant digitization, and plant processes in order to establish production systems that are more flexible and adaptable to external changes with high sustainability. The full integration of different support systems in the digital factory will strengthen communication across all R&D, production, marketing, and other organizational activities and thus facilitate customers to view the production of their products in real time and can suggests last minute modifications. This chapter presents the scenario of current manufacturing facilities which are still lacking in predictable maintenance, decision performance, early awareness, self-optimization and self-organizing features of the Industry 4.0. The Factories of the Future (Industry 4.0) will produce products in a smarter and more integrated, flexible and efficient way. Integrated sensors and IT systems can share and analyze data to predict failures, redesign, and trigger automatic repair processes and thus resulting new levels of performance to produce quality goods at reduced cost. The key technologies of the future industry such as virtual reality, simulation, augmented reality, cyber physical systems (CPS), artificial intelligence (AI), Internet of Things (IoT) and Industrial Internet of Things (IIoT), cloud computing, big data, and additive manufacturing have been highlighted in the chapter. In addition, the chapter also discussed state-of-the-art production technologies to meet quality improvement, reduce costs, and reduced lead time challenges owing to global competition, rapidly changing customer needs and low domestic productivity. At the end of chapter, socio-econo-techno justification of the Factories of the Future has been presented.

Corresponding author: tp_tiet@yahoo.co.in

Chandan Deep Singh and Harleen Kaur (eds.) Factories of the Future: Technological Advancements in the Manufacturing Industry, (1–20) © 2023 Scrivener Publishing LLC

Keywords: Industrial revolution, plant digitization, driving technologies, Internet of Things (IoT), smart factory, advanced manufacturing technologies

1.0   Introduction

Industrial change is the transition to advanced production strategies. These changes include shifting from manual manufacturing to machinery, production of new chemicals, increased use of unconventional production processes, development of machine tools and the growth of the digital industry system. Industries are currently aiming to move from mass production to customized production. The step towards entry into production strategies, which were completely different from the past, is called industrial revolution (Figure 1.1).

First Industrial Revolution
The First Industrial Revolution began in the 18th century and focused on the strength of steam and textile industries. During this time revolutionists from Europe and the United States built tools and machinery for the production of machinery. The original fibers were produced in simple spinning wheels, and a new mechanical version resulted in eight times higher production. Steam power was already known. Its use for industrial purposes was a great achievement for increasing human productivity. Steam engines could be used for weaving looms instead of man power. Developments such as steam locomotives brought about major changes because people and goods could travel long distances in just a few hours [2].

Second Industrial Revolution
The Second Industrial Revolution, which began in the 19th century, focused on the steel industry, the automotive industry, the production line, and the development of electricity. Henry Ford (1863-1947) introduced the concept of integration in the production of cars. In the assembly line, the complex work of assembling the many parts into a finished product was divided into

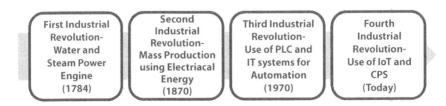

Figure 1.1 Industrial revolution [1].

a series of smaller tasks, resulting in higher productivity as each worker was required to assemble one or two parts in its place in the assembly line.

Third Industrial Revolution
The Third Industrial Revolution, also known as the "Digital Transformation" began in the 20th century and used electronic and information technology (IT) to produce production using programmable logic control (PLC) and computers. This technology is able to automate the entire production process without human intervention using programmed robots.

Fourth Industrial Revolution
Today, the Fourth Industrial Revolution, also known as "Industry 4.0" is based on the development of the Third Industrial Revolution. The Federal Government of Germany introduced Industry 4.0 as an emerging structure where production systems and goods in the form of Cyber Physical Systems (CPS) make extensive use of global information and communication network for automated information exchange for production and business procedures [3]. The four main drivers of Industrial 4.0 are Internet of Things (IoT), Industrial Internet of Things (IIoT), cloud-based production and intelligent production that helps transform the production process into a complete and intelligent digital [1, 4]. When these resources come together, Industry 4.0 has the potential to bring about dramatic improvements in the factory environment. Examples include machines that can predict failure and trigger automatic repair processes or self-programming that respond to unexpected changes in production [2].

1.1   Factory of the Future

The development of new technologies is making drastic changes in industrial production processes, resulting in "the factory of the future." The industry of the future is a vision of how producers should improve productivity by making progress in three phases: plant structure, plant digitization, and plant processes to establish production systems that are more flexible, adaptable to external changes and high sustainability [5, 6].

1.1.1   Plant Structure

The future factory plant structure has a flexible, multi-dimensional structure, with the setting of modular lines and environmentally sustainable production processes. The multidirectional structure employs driverless transport methods and is individually controlled by production in conjunction with production equipment. Such transport systems are guided by a laser scanner and the technology to detect radio frequency instead of a fixed transmitter thus making the integration structure fully flexible with flexible line modules. The future factory is designed for environmentally sustainable production, which combines energy efficiency with building materials for example enabling all LED lighting in the industry resulting in very low energy consumption.

1.1.2 Plant Digitization

Smart automation or Plant digitization can be done in a variety of ways as listed below for product development:

- Using robots that will ensure repetition and reproduction in complex tasks compared to workers. Robots can also collect information on each piece of work produced and automatically adjust their actions to their features. Robots can also support people in completing tasks in hard-to-reach areas.
- Using additive manufacturing or 3D printing, a computer-controlled process that creates three-dimensional objects by inserting objects, usually in layers. With additive manufacturing, design changes can be made quickly and efficiently during the production process with minimal or no damage.
- With augmented reality, such as smart mirrors, enables employees to see information as the overlay of their viewing field. This information is especially useful, for example, in assembling, maintenance and repairing things.
- Implementing simulations using real-time data, 3D production presentations to improve processes and flow of goods. 3D flow simulation simplifies dynamic responses to changes and allows operators to see work flow before adjusting the production line [5].
- Training methods have been developed that use 3D simulations to help staff learn in a real-world environment.

1.1.3 Plant Processes

Through the use of new digital technologies, manufacturers further develop their product designs and production processes continuously according to customer requirements.

1.1.4 Industry of the Future: A Fully Integrated Industry

Figure 1.2 shows the full integration of value chain with different support systems in the future factory. The value chain on the left side consists of suppliers, a manufacturing component, a press shop, a body shop, a paint store, a final assembly line and a customer, while, the support systems on the right side incorporate digital logistics, production simulation, and

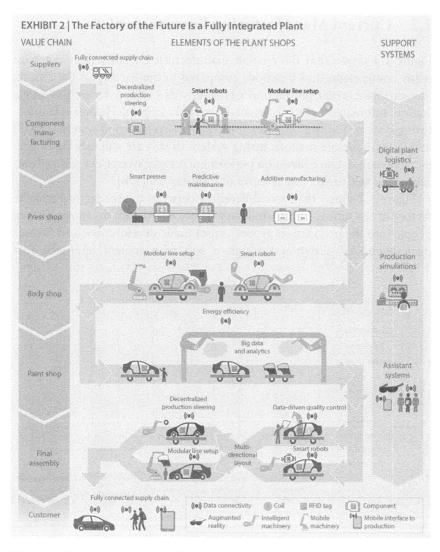

Figure 1.2 Factory of the future: Fully integrated plant [5].

various auxiliary programs. Throughout the value chain, production will be facilitated by the full integration of IT systems such as intelligent robots, modular line configurations, data-driven quality control etc. This integration will strengthen communication across all R&D, production, marketing, and other organizational activities. Customers will be able to view the production of their products (cars in this case) in real time and request changes at the last minute.

1.2   Current Manufacturing Environment

Figure 1.3 shows that the current manufacturing facility does not have many components and functions compared to the Factory of the Future. The various current production environments (Table 1.1) such as single station automated cells, Automated assembly system, Flexible manufacturing system (FMS), Computer-integrated manufacturing system (CIMS) and Reconfigurable manufacturing system (RMS) are still lacking in predictable maintenance, decision performance, early awareness, self-optimization and self-organizing features of Industry 4.0 [7, 8].

Figure 1.4 shows the basic differences between today's factory and the factory of the future. Today's industry is concerned with the integration of people into the production process, sustainable development and focuses on value-added activities through a soft management approach that

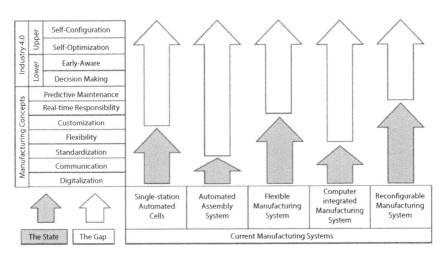

Figure 1.3 Technology gap between the current manufacturing system and Industry 4.0 [8].

Table 1.1 Current manufacturing environments.

Single station automated cells	A fully automatic production machine unattended for more than one cycle of operation. This system is an automated system that is simple and inexpensive to use with low labor costs and high productivity levels compared to the human-controlled cellular system.
Automated assembly system	The automated assembly system employs mechanical and automated devices to perform different assembly functions in an assembly line or cell. Automatic integration systems are usually designed to perform a consistent sequence of steps on a specific product that is produced in very large quantities.
Flexible manufacturing system (FMS)	A Flexible Manufacturing System (FMS) is a production system designed to easily adapt to changes in the type and quantity of a product being produced. Computer equipment and systems can be configured to produce various components and to adapt to changing production standards. Example: a NC machine, a pallet changer and a part buffer.
Computer integrated manufacturing system (CIMS)	CIMS refers to the use of computer-controlled equipment and automated production systems. CIMS integrates operations, marketing, design, development, production management, process management and other business processes to provide a seamless production process that reduces manual labor and creates repetitive tasks. The CIMS method accelerates the production process and uses real-time sensors and closed loop control systems to perform the production process automatically. It is widely used in the automotive, aviation, space and shipbuilding industries [9, 10].
Reconfigurable manufacturing system (RMS)	RMS has the ability to redesign hardware and manage resources at all operational and organizational levels, in order to quickly adjust production capacity and performance in response to sudden market changes or regulatory requirements [11].

	data source	Today's factory		Factory of the future	
		main characteristics	main technologies	main characteristics	main technologies
component	sensor	precision	smart sensors and failure detection	knowledge of own operations, predictive ability	monitoring of all features, life expectancy forecasts
machine	controller	manufacturability and performance	state-based system monitoring and diagnostics	knowledge of own operations, predictive ability, comparability ability	real-time preventive status indicators
manufacturing system	networked	performance and total asset efficiency	Lean operations: work and waste reduction	self-configuration, self-maintenance, self-organizing ability	risk exemption, performance

Figure 1.4 Comparison of today's factory and the factory of the future [14].

reduces complexity and cost by eliminating waste. Provides strategies to engage all employees in continuous review and improve efficiency.

However, the future factory (Industry 4.0) will produce products in a smarter and more integrated, flexible and efficient way. Integrated sensors and IT systems can share and analyze data to predict failures, redesign, and adapt to change. Individual segregation occurs in decision-making processes and enables real-time independent decisions at the machine level as well as flexible decisions regarding production processes based on timely data. Manufacturers can reach new levels of performance. They can, for example, move forward from prevention to predictable correction, which means that corrective actions are performed only when necessary. Better monitoring of products and production processes can also increase relationships with suppliers, produce quality goods and reduce costs [12, 13]. By increasing clarity, improving forecasting, and, finally, enabling automated systems, Industry 4.0 promotes faster, more flexible, and more efficient processes.

1.3   Driving Technologies and Market Readiness

Communication, automation, and optimization are the driving technologies of Industry 4.0 digital transformation. The key technologies of the future industry are discussed below:

(a) **Virtual Reality (VR)**: A virtual reality space where people can do things and participate in that environment using VR glasses. VR has been used in the design process to provide a more precise display and immersive creation of 3D models. VR has also been used in productive training programs, problem solving and remediation programs [7, 15, 16].

(b) **Simulation**: Simulation model demonstrates the performance of an existing or proposed system such as running an assembly line, production planning and scheduling. Simulation can also used to prepare the pre-production machine tool settings in a visible area without physical examination and thus resulted in saving time and money during testing of the production system.

(c) **Augmented Reality (AR)**: Augmented Reality (AR) technology integrates virtual reality with real world using multimedia, 3D-Modeling, Real-time Tracking, Intelligent Interaction, Sensor and more. AR uses computer-generated visual information, such as text, images, 3D models, music, video, etc., in the real world after imitation [17]. This integration of simulated computer simulations in real-world contexts helps to identify a product in an existing environment. The training of new staff and product testing by showing the various conditions in the developed area have been found to be effective and save time.

(d) **Cyber Physical Systems (CPS)**: A Cyber Physical System (CPS) or intelligent system is a computer system in which a machine is controlled or monitored by computer-based algorithms. In CPS, sensors, actuators etc. (physically) are closely integrated with computing, storage, communication and control systems (cyber) [1, 18]. CPS sensors are able to detect mechanical failures and automatically configure error correction actions. CPS is also used for the efficient use of each work station with the help of operational cycle time for that station [19]. Some of the major CPS features are listed below:

(1) Intelligent Grid: Cyber Physical Systems is used in the production, transmission, distribution and operation of power generation components, thereby providing dual and control communication between the power grid and users [20].

(2) Smart Transport Systems: CPS is used in the transport system to improve traffic management performance.

(3) Public Infrastructure Monitoring: Different CPS sensors are used for accurate and continuous monitoring of buildings, dams, and bridges etc.

(4) Aeronautic Applications: Cyber-Physical Systems used for aircraft inspection equipment, communications with Pilot, Structural Health Monitoring, In-flight testing, and aircraft maintenance etc.

(e) **Artificial Intelligence (AI)**: Artificial intelligence (AI) is the ability of a computer or computer-controlled robot to perform tasks that are normally performed by humans because it requires human ingenuity and judgment. The artificial intelligence system is able to self-determine, optimize and automatically respond to physical changes such as changing production schedules, suspension or operation of any machine units, automatic machine tools and automatic warning of uncontrolled conditions [7, 21, 22].

Examples of Artificial Intelligence:

- Production robots
- Self-driving cars
- Smart helpers
- Effective health care management
- Automatic investment
- Visible travel booking agent
- Social media monitoring, etc.

(f) **Cloud Computing**: Cloud computing means storing and accessing data and programs online instead of your computer's hard drive. Cloud computing brings a variety of computer services such as servers, storage, websites, network, software, statistics, and online intelligence ("cloud") to provide faster innovation, flexible resources, and scale economy. In the future industry, different plant machinery and devices are connected to the same cloud to share information with each other in digital production facilities [1, 23].

(g) **Big Data**: Big data is a combination of formal and informal data collected by organizations for information and use in machine learning projects, predictable modeling and other advanced mathematical applications.

Big data is usually seen with three V's [24]:

- Large *volume* of data in many areas;
- Wide *variety* of data types that are usually stored in large data systems; and
- The *velocity* at which the data is collected and processed.

Big data analysis helps in real-time intelligent production decisions by considering customer feedback and their own ideas on the products they use or intend to use, and from that knowledge, manufacturers focus on their product design to attract more and more customers.

(h) **Internet of Things (IoT) and Industrial Internet of Things (IIoT)**: IoT is used for common home applications such as starting a coffee machine with your phone, adjusting your air temperature, car tracking apps, and so on. Household items or everyday items connected to the internet and are therefore controllable from a distance.

IIoT refers to the IoT branch which focuses on the manufacturing and agricultural industry, which connects everything that is visible through the internet. This collaboration between each component ensures that production facilities run smoothly and at low cost. With the IIoT system, data and information flow is faster and efficient; staff can work safely and at high speed. IIoT also assists in production planning, predictable correction and error detection, improved human machine interaction, efficient use of resources.

(i) **Additive Manufacturing (AM)**: Additive manufacturing is a specific 3D printing process. This process creates layers in layers by inserting materials according to 3D digital design data. Additive manufacturing technologies such as selective laser melting (SLM), fused deposition method (FDM), and selective laser sintering (SLS) result in faster and more economical production [25]. AM is also employed in prototyping testing and design of parts/structures for low cost and customer satisfaction.

1.4 Connected Factory, Smart Factory, and Smart Manufacturing

A connected industry or intellectual industry is a manufacturing facility that uses digital technology to allow seamless sharing of information between people, machines, and sensors.

There are two main principles for allowing communication in the industry or factory. The first is to reach the right level of continuous production, self-improvement, and quality. This leads to higher profits. The second goal of the connected factory is to empower staff. The combination of control, visibility, and flexibility offered by the new digital solutions makes it possible for production workers to make further, impactful improvements. Figure 1.5 shows intelligent or integrated production including various digital technologies such as virtual reality, simulation, additive manufacturing, IoT, CPS, AI, cloud computing, etc. A smart or "connected" factory is one where almost every aspect of the factory is visible and available for analysis. Using data and updates, digital processes, and tools enable the entire organization, from management to shop floor staff reach to a new level of efficiency and profitability.

The connected industry uses consistent data distribution to adapt to the changing needs of intelligent production within an organization in a fully integrated and flexible system. Automatic workflow, real-time tracking and scheduling, as well as energy efficiency result in reduced costs and wastage [26].

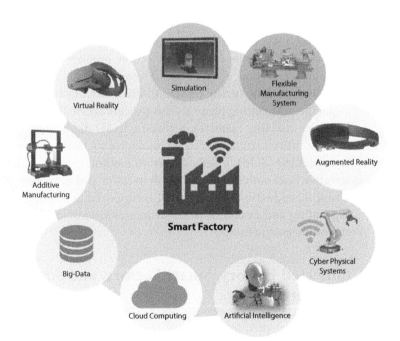

Figure 1.5 Connected industry or smart factory [7].

1.4.1   Potential Benefits of a Connected Factory

(a) High Productivity – Connected industries can perform jobs at a faster rate and run more efficiently leading to improved productivity and lower labor costs.

(b) Advanced Flexibility – Intelligent industries are designed for different production settings and demand flexibility. This provides complete flexibility of operation.

(c) Better Safety – The automation of tasks such as sorting, picking, packing, transporting, and delivering allows people to focus on safer jobs.

(d) Better Quality – The connected industry can detect quality problems quickly and can identify the cause.

(e) Lower Cost – More cost-effective processes, including asset management, better decisions, and improved service delivery.

(f) Flexible and effective communication as the factory floor and front and back offices share a single powerful data source [26, 27].

1.5   Digital and Virtual Factory

1.5.1   Digital Factory

Digital industry comprises network of digital models that replicate the features of a virtual factory. Digital Factory includes a list of methods and tools such as simulation and 3D visualization, all managed by integrated data management systems. The main goal of the digital industry is complete planning, continuous testing, and the development of a real productive factory [28]. The digital industry focuses on the following:

1. Improved quality of planning and economic efficiency
2. Shorter go-to market time
3. Clear communication
4. Similar planning standards
5. Managing competent information

What are the business benefits of a digital industry?
Complete and real-time data generated by digital factories promotes efficiency, productivity, safety and compliance. It also improves the flow

control of production work and the mobility of everything from immature items to continuous work and to finished goods. It also provides real-time access to operational data, so that managers can quickly overcome roadblocks and inefficiencies [29].

1.5.2   Virtual Factory

Virtual factory is based on integrated model that incorporates various software, tools, and solutions to any real-time production system problem. This model sees the real industry as a combination of various subsystems and integrates them. In practice, it creates a visual simulation work that helps to replicate the real life situation and that helps in design and implementation.

Virtual factory benefits include [30]:

- It assists in building skills to support rapid development in the manufacturing sector by bringing together professionals.
- It helps to provide solutions in a fast and inexpensive way.
- Eliminates the need for testing or pilot studies and replaces it with virtual simulation through software.
- It helps in optimum decision making.

1.6   Advanced Manufacturing Technologies

Manufacturing organizations use state-of-the-art production technology to meet quality improvement, reduce costs, and reduced lead time challenges posed by global competition, rapidly changing customer needs and low domestic productivity [31]. Advanced production technologies include:

- **Computer Technology (e.g., CAD, CAE, CAM)** – CAD, computer-assisted design, computer use to design 2D and 3D models. CAD provides a preview of the final product with digital visibility of the final product and its components and thus improves the quality of the design with greater accuracy and minor errors. Computer-assisted engineering (CAE) is the use of computer software to simulate performance in order to improve product designs or to help solve engineering problems in many industries. This includes the imitation, validation and efficiency of products, processes, and production tools. CAM, a computer-assisted production, provides an intelligent way to generate code using software such as GUIs (Graphical User Interfaces) to control the equipment involved in the production process.

- **High-Performance Computing (HPC)** – HPC uses information communication technology in all production and transportation systems. This program focuses on the creation and implementation of programs to improve production facilities with effective monitoring that is in line with the constantly updated and remedied plans.

- **Rapid prototyping (additive manufacturing)** – 3D Printing or Rapid prototyping is the rapid construction of a visual component, model or combination using computer-assisted design (CAD). Rapid Prototyping helps designers present new ideas to board members, clients or investors to understand and approve a development or product. These displays can also allow designers to get the right feedback from customers based on the actual body product rather than the concept.

- **High precision technology** – such as NC (Numerical Controller), CNC Machine (Computer Numerical Controller) increases production capacity and reduces set switching time.

- **Network and IT integration** – With internet access to all aspects of production, there is an instant notification of any potential problems, allowing blockchain adjustment and saving time and money.

- **Advanced robots and other intelligent production systems** – Designed to automate processes that include precise movement, lifting heavy objects, and consistent integration of features into production systems. Additionally, robots minimize the risks involved in hazardous activities such as the automotive industry and aerospace.

- **Automated technology** – Integrates all the processes and equipment that make plants and systems work automatically. These include Interactive Robots, Artificial Intelligence (AI), Internet of Things (IoT), etc. automatically performs work procedures correctly and with a negligible error rate.

- **Monitoring systems and Control systems** – Monitoring systems such as sensor used to view and record industrial process data. However, control systems continuously maintain or change the state of the system through actuators. These control systems help maintain quality, yield and energy efficiency and ensure that operations are carried out safely and profitably.

- **New industrial field technologies (e.g., composite materials)** – Advanced materials enable more precise integration of specific applications such as metal, plastic, glass, and ceramics. Materials that require precision in its chemical and physical properties are used to empower business success.

1.6.1   Advantages of Advanced Manufacturing Technologies

The benefits of using state-of-the-art production technology are given below:

1. Improved Quality Standards
   The potential benefit of advanced production technology is quality improvement. Switching to robots and automation in the production process leads to almost zero human error. The number of accidents, errors, and resulting cost inefficiencies are also reduced.
2. Improved Production
   Advanced production technology improves productivity in many ways. It helps producers to increase or decrease depending on market demand. From creating custom products in small batches to large-scale production, productivity is improved and can be customized as well.
3. Encourages Innovation
   Being able to measure productivity gives manufacturers the ability to create new products in a less expensive way. A small, personalized product can be created without compromising on normal production times.
4. Reduced Production Time
   Digital production uses virtualization to create digital industries that mimic the production process. Such simulation helps engineers design a seamless factory structure, production sequence, and model output. Any potential obstacles can be resolved before production can begin.

1.7   Role of Factories of the Future (FoF) in Manufacturing Performance

Production process improvement is one of the most effective ways to increase quality, efficiency, and foundation. Developing processes that contribute to the final product is an effective way to impart scalable and sustainable changes. Proper upgrades can reduce errors, reduce production time, and increase customer satisfaction. The role of Factories of the Future in improving production performance is highlighted below:

- FOF promotes the productivity and efficiency of the organization, better flexibility and profits. FoF also improves customer self-awareness.
- FOF technology allows for faster and faster production while cost-effective and efficient distribution of resources.
- Intelligent technology improves automation, machine-to-machine communication, and decision-making.
- FoF provides fast batch switching, automatic tracking, tracing and reporting for better productivity.
- FOF plays an important role in information sharing and collaboration that allows production lines, business processes, and departments to communicate regardless of location, time zone, field, or anything else.
- FOF uses automatic tracking and tracking capabilities to resolve problems quickly leading to improved customer service and information.

1.8   Socio-Econo-Techno Justification of Factories of the Future

Industry 4.0 or Factories of the Future (FoF) is changing the industry in the way we know it since the industrial revolution. Imagine you have personalized products and services, fully customized for you. With FoF, this becomes a reality because it drives new inventions and introduces new technologies to such an extent that it will affect our entire society [32]. The following points illustrate the ways in which FoF will impact on society-economy-technology:

1. New production processes will make it possible to make products the way you want at the new level. You will be able to customize the products according to your needs at a detailed level, covering everything from cars to personal medicine.
2. Industry 4.0, especially when combined with machine learning and practical skills, will drastically change the working conditions of workers. Many jobs will disappear while we get more new jobs, and most duplicate jobs will go from manual to automation.
3. The high growth in demand for information and communication technology (ITC) by industrial organizations can build the future of Industrial 4.0 and may have positive effects on the various sectors of the firm [33].
4. New solutions can reduce energy consumption and thus help organizations to strengthen their business with existing and new business models.
5. FoF makes the world more digital, more connected, more flexible, and more responsive. Well-known public relations are changing beyond recognition; from business relationships to consumer to peer-to-peer approaches [34].
6. Industry 4.0 introduces new opportunities for health care, the ability to empower more people around the world to become entrepreneurs, and increased access to education [35].

In the fourth industrial revolution, the social impact of technological changes in the economic sphere, labor market, and innovation is better understood now than during the previous industrial revolution. Meanwhile, governments and policymakers need to adapt and respond quickly to the immediate emergence of the Industrial 4.0 landscape by providing an environment and policies that can guide the future of sustainable economic and social development and implement the technologies for Industry 4.0 on behalf of individuals and communities [33, 36].

References

1. Vaidya, S., Ambad, P., Bhosle, S., Industry 4.0 – A glimpse. *Proc. Manuf.*, 20, 233–238, 2018.

2. See https://www.desouttertools.com/industry-4-0/news/503/industrial-revolution-from-industry-1-0-to-industry-4-0.

3. Kamarul Bahrin, M.A., Othman, M.F., Nor Azli, N.H., Talib, M.F., Industry 4.0: A review on industrial automation and robotic. *J. Teknol.*, 78, 6-13, 137–143, 2016.

4. Erol, S., Jäger, A., Hold, P., Ott, K., Sihn, W., Tangible Industry 4.0: A scenario-based approach to learning for the future of production. *Proc. CIRP*, 54, 13–18, 2016.

5. See https://www.bcg.com/publications/2016/leaning-manufacturing-operations-factory-of-future.

6. de Asis Marti Nieto, F., Goepp, V., Caillaud, E., From factory of the future to future of the factory: Integration approaches. *IFAC-PapersOnLine*, 50, 1, 11695–11700, 2017.

7. Phuyal, S., Bista, D., Bista, R., Challenges, opportunities and future directions of smart manufacturing: A state of art review. *Sustainable Futures*, 2, 100023, 2020.

8. Qin, J., Liu, Y., Grosvenor, R., A categorical framework of manufacturing for Industry 4.0 and beyond. *Proc. CIRP*, 52, 173–178, 2016.

9. See https://www.techopedia.com/definition/30965/computer-integrated-manufacturing-cim.

10. Sheng, Y.K. and Lan, W.L., Research on computer integrated manufacturing system based on integrated logistics. *Adv. Mat. Res.*, 271–273, 738–741, 2011.

11. Bi, Z.M., Lang, S.Y.T., Shen, W., Wang, L., Reconfigurable manufacturing systems: The state of the art. *Int. J. Prod. Res.*, 46, 4, 967–992, 2008.

12. Nagy, J., Industry 4.0: Definition, elements and effect on corporate value chain, Workshop studies - Institute of Business Economics, Corvinus University of Budapest, Hungary, 2017. http://unipub.lib.uni-corvinus.hu/3115/.

13. Skapinyecz, R., Illés, B., Bányai, Á., Logistic aspects of Industry 4.0. *IOP Conf. Ser.: Mater. Sci. Eng.*, 448, 012014, 2018.

14. Lee, J., Bagheri, B., Kao, H.A., Recent advances and trends of cyber-physical systems and big data analytics in industrial informatics. *Proc. Int. Conf. on Industrial Informatics (Porto Alegre)*, IEEE, 2014.

15. Salah, B., Abidi, M.H., Mian, S.H., Krid, M., Alkhalefah, H., Abdo, A., Virtual reality based engineering education to enhance manufacturing sustainability in Industry 4.0. *Sustainability*, 11, 5, 1477, 2019.

16. Berg, L.P. and Vance, J.M., Industry use of virtual reality in product design and manufacturing: A survey. *Virtual Real.*, 21, 1, 1–17, 2017.

17. Chen, Y., Wang, Q., Chen, H., Song, X., Tang, H., Tian, M., An overview of augmented reality technology. *J. Phys.: Conf. Ser.*, 1237, 022082, 2019.

18. Bagheri, B., Yang, S., Kao, H.A., Lee, J., Cyber-physical systems architecture for self-aware machines in Industry 4.0 environment. *IFAC Conf.*, 38, 30, 1622–1627, 2015.

19. Kolberg, D. and Zühlke, D., Lean automation enabled by Industry 4.0 technologies. *IFAC Conf.*, 38, 3, 1870–1875, 2015.

20. Bhrugubanda, M., A review on applications of cyber physical systems. *Int. J. Innov. Sci. Eng. Technol.*, 2, 6, 728–730, 2015.

21. Lee, J., Davari, H., Singh, J., Pandhare, V., Industrial artificial intelligence for Industry 4.0-based manufacturing systems. *Manuf. Lett.*, 18, 20–23, 2018.

22. Genge, B., Nai Fovino, I., Siaterlis, C., Masera, M., *Analyzing cyber-physical attacks on networked industrial control systems*, pp. 167–183, Springer Berlin Heidelberg, Berlin, 2011.

23. Marilungo, E., Papetti, A., Germani, M., Peruzzini, M., From PSS to CPS design: A real industrial use case toward Industry 4.0, The 9th CIRP IPSS Conference: Circular perspectives on product/service-systems. *Proc. CIRP*, 64, 357–362, 2017.

24. See https://searchdatamanagement.techtarget.com/definition/big-data.

25. Landherr, M., Schneider, U., Bauernhansl, T., The Application Centre Industrie 4.0 - Industry-driven manufacturing, research and development, 49th CIRP Conference on Manufacturing Systems (CIRP-CMS). *Proc. CIRP*, 57, 26–31, 2016.

26. See https://ottomotors.com/blog/connected-factory-manufacturing.

27. Depince, P., Chablat, D., Noel, E., Woelk, P.O., The virtual manufacturing concept: Scope, socio-economic aspects and future trends. *Proceedings of DETC'2004, ASME Design Engineering Technical Conferences and Computers and Information in Engineering Conferences*, 2004.

28. See https://www.tibco.com/reference-center/what-is-a-digital-factory.

29. See https://www.cognizant.com/us/en/glossary/virtual-reality.

30. See https://www.mbaskool.com/business-concepts/operations-logistics-supply-chain-terms/13306-virtual-factory.html#google_vignette.

31. Saraph, J.V. and Sebastian, R.J., Human resource strategies for effective introduction of Advanced Manufacturing Technologies (AMT). *Prod. Inventory Manage. J.*, 33, 1, 64, 1992.

32. See https://it-voices.com/en/articles-en/5-life-changing-ways-industry-4-0-will-change-society.

33. Morrar, R., Arman, H., Mousa, S., The Fourth Industrial Revolution (Industry 4.0): A social innovation perspective. *Technol. Innov. Manage. Rev.*, 7, 11, 12–20, 2017.

34. Arroyo, L., Murillo, D., Val, E., *Trustful and trustworthy: Manufacturing trust in the digital era*, ESADE Roman Llull University Institute for Social Innovation; EY Fundacion Espana, Barcelona, 2017.

35. Schreiber, U., EY women. Fast forward. An interview with Uschi Schreiber. *Leaders*, 39, 1, 40, 2017.

36. Schwab, K., *The Fourth Industrial Revolution*, World Economic Forum, Geneva, 2015.

Industry 5.0

**Talwinder Singh[1], Davinder Singh[1], Chandan Deep Singh[1]
and Kanwaljit Singh[2*]**

*[1]Department of Mechanical Engineering, Punjabi University,
Patiala, Punjab, India
[2]Department of Mechanical Engineering, Guru Kashi University,
Talwandi Sabo, Punjab, India*

Abstract

The world has seen a massive increase in environmental pollution beginning from the Second Industrial Revolution. However, unlike in the past several decades, the manufacturing industry is now more focused on controlling different aspects of waste generation and management and on reducing adverse impacts on the environment from its operation. Having environmental awareness is often considered a competitive edge due to the vast amount of support from government; international organizations like the UN, WHO, etc.; and even an ever-growing niche customer base that supports environmentally friendly companics. Unfortunately, Industry 4.0 does not have a strong focus on environmental protection, nor has it focused technologies to improve the environmental sustainability of the Earth, even though many different AI algorithms have been used to investigate from the perspective of sustainability in the last decade. While the existing studies linking AI algorithms with environmental management have paved the way, the lack of strong focus and action leads to the need for a better technological solution to save the environment and increase sustainability. We envisage this solution to come out of Industry 5.0. Further, the objective is to develop an understanding about Industry 5.0 along with the sustainability of Industries and their performance at a time when they are adopting advanced production techniques and while adopting these, they have to keep in mind environmental aspects. Further, it involves knowledge regarding improving performance of business firms while going for technical advancements and thus gaining in competition. The term Industry 5.0 refers to people working

*Corresponding author: kanwalpatiala05@gmail.com

Chandan Deep Singh and Harleen Kaur (eds.) Factories of the Future: Technological Advancements in the Manufacturing Industry, (21–46) © 2023 Scrivener Publishing LLC

alongside robots and smart machines. It's about robots helping humans work better and faster by leveraging advanced technologies like the Internet of Things (IoT) and big data. It adds a personal human touch to the Industry 4.0 pillars of automation and efficiency. In manufacturing environments, robots have historically performed dangerous, monotonous or physically demanding work, such as welding and painting in car factories and loading and unloading heavy materials in warehouses. As machines in the workplace get smarter and more connected, Industry 5.0 is aimed at merging those cognitive computing capabilities with human intelligence and resourcefulness in collaborative operations. Since the use cases of Industry 5.0 are still in their relative infancy, manufacturers should be actively strategizing ways to integrate human and machine workers in order to maximize the unique benefits that can be reaped as the movement continues to evolve.

Keywords: Industry 4.0, Industry 5.0, block chain, fog computing, exoskeleton, IoT, AI, manufacturing

2.1   Introduction

Industry 4.0 came with promises of sustainable and efficient manufacturing. But it has been criticized for being unable to deliver due to its focus on mass production rather than sustainability. However, sustainability and human wellbeing lies at the heart of what comes next--Industry 5.0. The first Industrial revolution brought mechanical power, the second electrical energy, and the third automation [1]. In the fourth industrial revolution, the Internet of Things (IoT), cloud computing, and machine-to-machine learning was used for increased automation and better communication. Machines began to operate without the need for human intervention.

2.1.1   Industry 5.0 for Manufacturing

The Fourth Industrial Revolution changed manufacturers from being physical systems to a mix between cyber and physical systems. The concept behind Industry 4.0 revolves around cyber and physical systems communicating with each other via the IoT. This communication and the subsequent manipulation of data allow manufacturers to become adaptive, intelligent, and flexible [2]. The main drivers of Industry 4.0 can be listed as follows:

The Internet and IoT being available almost everywhere;

- Business and manufacturing integration;
- Digital twins of real-world applications;
- Efficient production lines and smart products.

By embracing Industry 4.0, manufacturers have decreased:

- Production costs by 10–30%
- Logistic costs by 10–30%
- Quality management costs by 10–20%

2.1.1.1   Industrial Revolutions

(a) **Industry 1.0** The methodology of industry 5.0 evolved from the late eighteenth century when water and steam-powered machines were made for mass production of goods thereby removing the barrier of serving limited number of customers and lead to the expansion of business and was referred as the era of Industry 1.0

(b) **Industry 2.0** The starting of the twentieth century gave rise to the next industrial upheaval, which came into limelight because of the production of machines running on electricity. These machines not only reduced the human effort but they were also easy to operate and were better than machines operated on steam and water as they were not resource hungry

(c) **Industry 3.0** With the more advancements in the electronic industry, automation in production and manufacturing sector promoted to a new level as many new electrical devices came into existence such as transistors, PLC's and integrated circuits automated the machines which substantially reduced human labor, increased pace, improved accuracy and in fact full substitution of human body. This level of automation was referred to as Industry 3.0

(d) **Industry 4.0** The blast in the web and media transmission industry in the late 1990s took the automation to a new level when the connection and exchange of information were done in totally different ways as compared with earlier ones. Cyber Physical System was one property that enabled us to merge the physical world with the virtual one where machines became smarter enough to communicate with each other without any physical or geographical barrier and is known as Industry 4.0

(e) **Industry 5.0** This era of industrial upheaval attempts to take the Industry 4.0 to a new milestone by integrating humans and machines in the smart factory. It will also enable the implementation of critical factors such as mass customization and personalization. It will also enhance the skills of the workers and improve their knowledge about the manufacturing processes

The fourth industrial revolution is based on the idea of merging the physical and virtual worlds through cyber-physical systems, and interconnecting humans, machines and devices through the Internet of Things. This horizontal and vertical interconnection across entire value chains, from customer to supplier, across the entire product lifecycle, and across different functional departments forms new value networks and ecosystems. The creation of added value can be made more efficient, personalized, of higher quality, service-oriented, traceable, resilient and flexible. Maintenance will be connected to production in a different way than before, resulting in an integrated chain that covers the entire life cycle. It will generate benefits in economic, ecological and social regards, relating to the Triple Bottom Line of sustainable development [3].

The four industrial revolutions are centered around general-purpose technologies, in the first industrial revolution mechanization through water and steam power, electrification, labor division and mass production in the second industrial revolution, and IT, electronics and automation in the third industrial revolution. Industry 4.0 focuses on cyber-physical-systems and the Internet of Things, as well as further technologies mentioned when referring to the concept. Instead, Industry 5.0 shall base on supporting and fostering socially and ecologically relevant values.

Industry 5.0 recognizes the power of industry to achieve societal goals beyond jobs and growth to become a provider of prosperity, by making production respect the boundaries of our planet and placing the wellbeing of the industry worker at the center of the production process [4].

However, the methods used to measure industrial output have not changed since the first Industrial Revolution. Along with a turn towards a more customizable and personalized human-centered model, which we covered in this blog, a second key tenet of Industry 5.0 is sustainability [5].

For manufacturing to be sustainable it must consider intangible measurements relating to:

- The environment
- Society
- Fundamental human rights

Industry 5.0 harnesses the benefits of Industry 4.0 but extends it by taking a human-centered approach. Recovery from the current pandemic crisis relies on changes rapidly being made in the environmental and digital spheres – so we can build an economy that is both sustainable and resilient.

Industry 5.0 is made possible through deliberately focusing on research and innovation as well as putting technology at the forefront of the transition. It is characterized as being defined by a purposefulness that is more than just manufacturing goods for profit. The three central tenets of Industry 5.0 are: human-centricity, sustainability, and resilience.

The benefits of Industry 5.0 for sustainable manufacturing

- Manufacturers can become more competitive by cutting manufacturing costs.
- Manufacturers, workers, and society are the net winners.
- Workers are empowered through being put at the center of industry through up-skilling and re-skilling.
- Manufacturers can focus on solutions including conserving resources, tackling climate change, and promoting social stability while continuing to be profitable.
- Sustainable manufacturing is promoted through circular production models, supported by advanced technologies that efficiently use resources.
- Focusing on sustainability, industry becomes more resilient against external shocks, such as Brexit, pandemics, and financial crises [6].

2.1.2   Real Personalization in Industry 5.0

Previous industrial revolutions demonstrate that manufacturing systems and strategies have been continuously changing towards greater productivity and efficiency. Although many conferences and symposia are being held with a focus on Industry 5.0, there are still several manufacturers and industry leaders under the belief that it is too soon for a new industrial revolution. On the other hand, accepting the next industrial revolution requires the adoption, standardization, and implementation of new technologies, which needs its own infrastructure and developments.

Industry 5.0 will bring unprecedented challenges in the field of human–machine interaction (HMI) as it will put machines very close to the everyday life of any human. Even though we are obsessed with machines such as programmable assistive devices and programmable cars, we do not consider them a version of cobots (even though the differences are not that great from a certain perspective), mostly because of their shape. Cobots will be very different as their organization and introduction will contain human-like functionalities such as gripping, pinching, and interaction based on intention and environmental factors [7]. We also anticipate that Industry 5.0 will create many jobs in the field of HMI and computational human factors (HCF) analysis.

Industry 5.0 will revolutionize manufacturing systems across the globe by taking away dull, dirty, and repetitive tasks from human workers wherever possible. Intelligent robots and systems will penetrate the manufacturing supply chains and production shopfloors to an unprecedented level. This will be made possible by the introduction of cheaper and highly capable robots, made up of advanced materials such as carbon fiber and lightweight but strong materials, powered by highly optimized battery packs, cyber attack hardened, with stronger data handling processes (i.e., big data and artificial intelligence), and a network of intelligent sensors.

Industry 5.0 will increase productivity and operational efficiency, be environmentally friendly, reduce work injury, and shorten production time cycles. However, contrary to immediate intuition, Industry 5.0 will create more jobs than it takes away. A large number of jobs will be created in the intelligent systems arena, AI and robotics programming, maintenance, training, scheduling, repurposing, and invention of a new breed of manufacturing robots. In addition, since repetitive tasks need not be performed by a human worker, it will allow for creativity in the work process to be boosted by encouraging everyone to innovatively use different forms of robots in the workplace [8].

Furthermore, as a direct impact of Industry 5.0, a large number of start-up companies will build a new ecosystem of providing custom robotic solutions, in terms of both hardware and software, across the globe. This will further boost the global economy and increase cash flow across the globe.

Bio-inspired technologies and processes stemming from the concept of Biological Transformation can be integrated with, for instance, the following properties:

- ☐ Self-healing or self-repairing
- ☐ Lightweight
- ☐ Recyclable
- ☐ Raw material generation from waste
- ☐ Integration of living materials
- ☐ Embedded sensor technologies and biosensors
- ☐ Adaptive/responsive ergonomics and surface properties
- ☐ Materials with intrinsic traceability

Digital twins and simulation technologies optimize production, test products and processes and detect possible harmful effects, for instance:

- ☐ Digital twins of products and processes
- ☐ Virtual simulation and testing of products and processes (e.g., for human-centricity, working and operational safety)
- ☐ Multi-scale dynamic modelling and simulation
- ☐ Simulation and measurement of environmental and social impact
- ☐ Cyber-physical systems and digital twins of entire systems
- ☐ Planned maintenance

Energy-efficient and secure data transmission, storage, and analysis technologies are required, with properties such as:

- ☐ Networked sensors
- ☐ Data and system interoperability
- ☐ Scalable, multi-level cyber security
- ☐ Cyber security/safe cloud IT-infrastructure
- ☐ Big data management
- ☐ Traceability (e.g., data origin and fulfilment of specifications)
- ☐ Data processing for learning processes
- ☐ Edge computing

Artificial intelligence, nowadays often still referring to advanced correlation analysis technologies, must be developed further in several regards:

- ☐ Causality-based and not only correlation-based artificial intelligence
- ☐ Show relations and network effects outside of correlations
- ☐ Ability to respond to new or unexpected conditions without human support

☐ Swarm intelligence
☐ Brain-machine interfaces
☐ Individual, person-centric Artificial Intelligence
☐ Informed deep learning (expert knowledge combined with Artificial Intelligence)
☐ Skill matching of humans and tasks
☐ Secure and energy-efficient Artificial Intelligence
☐ Ability to handle and find correlations among complex, interrelated data of different origin and scales in dynamic systems within a system of systems

As the majority of technologies mentioned require large amounts of energy to operate, the following technologies and properties are required to achieve emission neutrality:

☐ Integration of renewable energy sources
☐ Support of Hydrogen and Power-to-X technologies
☐ Smart dust and energy-autonomous sensors
☐ Low energy data transmission and data analysis

2.1.3 Industry 5.0 for Human Workers

A human-centered approach prioritizes human needs over the production process. Manufacturers must identify what technology can do for the workers, and address how technology can adapt to the needs of the worker rather than the other way around. It is important that technology does not affect issues such as privacy and autonomy.

For manufacturing to be sustainable, it must develop circular processes that reuse, repurpose, and recycle resources. Environmental impacts need to be reduced. Sustainable manufacturers can harness the power of technologies such as AI and additive manufacturing to increase personalization, which optimize resource-efficiency and minimize waste. Manufacturers must develop a higher degree of robustness in industrial production to better protect themselves against disruptions and crises such as covid-19.

Industry 4.0 is about automating processes and introducing edge computing in a distributed and intelligent manner. Its sole focus is to improve the efficiency of the process, and it thereby inadvertently ignores the human cost resulting from the optimization of processes. This is the biggest problem that will be evident in a few years when the full effect of Industry 4.0 comes into play. Consequently, it will face resistance from labor unions and politicians, which will see some of the benefits of Industry 4.0 neutralized as pressure to improve the employment number increases [9].

However, it is not really necessary to be on the back foot when it comes to introducing process efficiency by means of introducing advanced technologies. It is proposed that Industry 5.0 is the solution we will need to achieve this once the backward push begins.

The enabling technologies of Industry 5.0 are a set of complex systems that combine technologies, such as smart materials, with embedded, bio-inspired sensors. Therefore, each of the following categories can only unfold its potential when combined with others, as a part of systems and technological frameworks [10].

2.2   Individualized Human-Machine-Interaction

To link humans with technologies, support humans and combine human innovation and machines capabilities. The following technologies support humans in physical and cognitive tasks:

- ☐ Multi-lingual speech and gesture recognition and human intention prediction
- ☐ Tracking technologies for mental and physical strain and stress of employees
- ☐ Robotics: Collaborative robots ('cobots'), which work together with humans and assist humans
- ☐ Augmented, virtual or mixed reality technologies, especially for training and inclusiveness
- ☐ Enhancing physical human capabilities: Exoskeletons, bio-inspired working gear and safety equipment
- ☐ Enhancing cognitive human capabilities: Technologies for matching the strengths of Artificial Intelligence and the human brain (e.g., combining creativity with analytical skills), decision support systems

First, if Industry 5.0 covers the technologies relevant for Industry 4.0, this could lead to a potential confusion when coining a result or development associated with the fourth industrial revolution as a fifth industrial revolution. Previous industrial revolutions' developments took decades to unfold, while Industry 4.0 was first described in 2011. Industry 4.0 as a concept is already described as rather marketing-driven, so using the nomenclature once again after such a short time could amplify this perception. For instance, a truly technology-driven development in the field of biological transformation or quantum technologies4 would be a separate industrial revolution, posing the question if this would then not be centered around societal and ecological values.

Second, creating the concept of Industry 5.0 around societal and ecological values instead of technologies, disrupting the concept of industrial revolutions in general, could lead to a misperception. This could drive the developments from a technological capability perspective to a political one. The concept of values and, more importantly, which values are seen as important and how they are understood, are perceived in several different ways around the world.

Third, many thoughts and concepts of Industry 5.0 can already be found in Industry 4.0. For instance, in its initial concept, Industry 4.0 was centered around values for humans, society, and ecology. Further, highly individualized products, often referred to as 'batch size one' as a catchphrase of mass customization, represent a central characteristic of Industry 4.0.5 Since the introduction of the term, several technologies have been labelled 'Industry 4.0' without referring to a broader purpose outside economic benefits. Nevertheless, technologies allowing human-machine-interaction such as Augmented and Virtual Reality and collaborative robotics, are part of the concept of Industry 4.0, and used across continents to support humans and generate value.

Fourth, and this is particularly relevant, Industry 4.0 is still unfolding to a large extent. Especially small and medium-sized enterprises, craft manufacturers, or traditional industries are still far from a wide-scale implementation of Industry 4.0 technologies. Further, while several technologies can be seen in isolated solutions, the full horizontal and vertical integration across the supply chain, as aimed for in Industry 4.0, is still far from reality in most industrial value chains. Therefore, the nomenclature of Industry 5.0 might indicate an even newer set of technologies than in Industry 4.0, but several of Industry 4.0 ideas are even further far away from implementation than proposed for Industry 5.0 [11].

To summarize, several of the ideas of Industry 4.0 seem to be revitalized under a new terminology. The concept of Industry 5.0 could also be described as re-introducing the lost dimension of a 'human/value-centered Industry 4.0', or as one participant put it, 'Industry 4.1'. At the end of the two workshops, however, there was a clear consensus that there is a need for adding a value dimension to the concept of Industry 4.0.

Bringing back human workers to the factory floors, the Fifth Industrial Revolution will pair human and machine to further utilize human brainpower and creativity to increase process efficiency by combining workflows with intelligent systems. While the main concern in Industry 4.0 is about automation, Industry 5.0 will be a synergy between humans and autonomous machines. The autonomous workforce will be perceptive and informed about human intention and desire. The human race will work alongside robots, not only with no fear but also with peace of mind, knowing that their robotic co-workers adequately understand them and have the ability to effectively collaborate with them. It will result in an exceptionally efficient and value-added production process, flourishing trusted autonomy, and reduced waste and associated costs.

Industry 5.0 will change the definition of the word "robot". Robots will not be only a programmable machine that can perform repetitive tasks but also will transform into an ideal human companion for some scenarios. Providing robotic productions with the human touch, the next industrial revolution will introduce the next generation of robot, commonly termed as cobots, which will already know, or quickly learn, what to do. These collaborative robots will be aware of the human presence; therefore, they will take care of the safety and risk criteria. They can notice, understand, and feel not only the human being but also the goals and expectations of a human operator. Just like an apprentice, cobots will watch and learn how an individual performs a task. Once they have learned, the cobots will execute the desired tasks as their human operators do. Therefore, the human experiences a different feeling of satisfaction while working alongside cobots [12].

2.3   Industry 5.0 is Designed to Empower Humans, Not to Replace Them

"Excessive automation at Tesla was a mistake. To be precise, my mistake." Elon Musk states the problems with Industry 4.0 in no uncertain terms. Robots replaced humans in droves, but industry is beginning to realize that it has made an error. The obsession with focusing on mass production backfired.

While robots can complete repetitive tasks far more consistently and precisely than humans, they are unable to problem-solve and intuitively address issues. This is invaluable in manufacturing where making judgements is a key component to ensuring the correct functioning of the whole system.

Industry 5.0 makes the shift from robots to cobots – robots that collaborate with humans, who are at the center of the process. Humans can use robots to carry out repetitive tasks such as tightening screws while they are free to critically think about the bigger picture. Human creativity is needed in Industry 5.0--problems can be solved by humans and fixed by robots.

2.4   Concerns in Industry 5.0

In the next industrial revolution, humans are expected to add high-value tasks in manufacturing policies. Standardization and legalization will help to prevent any serious issues between technology, society, and businesses.

- Particularly, senior members of a society and stakeholders will find it much more difficult to adapt with the new industrial revolution.
- Fast and highly efficient manufacturing may result in an overproduction phenomenon. Implementation transparency should also be taken into consideration.
- We must consider how autonomous systems can incorporate ethical principles.
- There should be explainable ethical behavior solutions in autonomous systems.
- Ethical behavior in autonomous systems must be subject to verification and validation.
- Essential skill gaps such as CROs in future management and executive roles must be addressed.

The Fifth Industrial Revolution will emerge when its three major elements – intelligent devices, intelligent systems, and intelligent automation – fully merge with the physical world in co-operation with human intelligence. The term "automation" describes autonomous robots as intelligent agents collaborating with humans at the same time, in the same workspace. Trust and reliability between these two parties will achieve promising efficiency, flawless production, minimum waste, and customizable manufacturing. In doing so, it will bring more people back to the workplace and improve the process efficiency. Manufacturers must advance cautiously. The potential pitfalls of excessive automation have been well documented and can lead to the inadvertent creation of new social and political structures. The challenge for Industry 5.0 is to get the balance right. Manufacturers should balance their needs against the expected outcomes. How can technologies and machines be used to benefit humans and the planet? Asking this question is the difference between Industry 5.0

and the previous revolutions – it symbolizes the shift in our relationship with technology and the planet [13].

The technologies supporting the concept of Industry 5.0 include:

- ☐ Human-centric solutions and human-machine-interaction technologies that interconnect and combine the strengths of humans and machines
- ☐ Bio-inspired technologies and smart materials that allow materials with embedded sensors and enhanced features while being recyclable
- ☐ Real time based digital twins and simulation to model entire systems
- ☐ Cyber safe data transmission, storage, and analysis technologies that are able to handle data and system interoperability
- ☐ Artificial Intelligence e.g. to detect causalities in complex, dynamic systems, leading to actionable intelligence
- ☐ Technologies for energy efficiency and trustworthy autonomy as the above-named technologies will require large amounts of energy.

For a systemic approach, several challenges must be regarded that can be addressed with respective enablers:

- ☐ In the social dimension, a human-centric approach needs to be developed into a socio-centric approach, addressing contemporary challenges, heterogeneous needs while integrating participation of the society to increase trust and acceptance.
- ☐ The high speed of transformation requires measures in the governmental and political dimension. This includes 33agile government approaches, understanding complex, interrelated systems of industrial ecosystems and labor markets
- ☐ Interdisciplinarity and transdisciplinarity, the requirement to integrate different research disciplines (e.g., life sciences, engineering, social sciences and humanities) is complex and must be understood in a systems approach.
- ☐ In the economic dimension, solutions for maintaining economic profitability and competitiveness and the funds required must be found, e.g., through developing respective business models that value ecological and social aspects.

☐ Scalability in the sense of ensuring a broad-scale implementation of technologies across value chains and ecosystems, including SMEs.

As such, the concept of Industry 5.0 is not based on technologies, but is centered around values, such as human-centricity, ecological or social benefits. This paradigm shift is based on the idea that technologies can be shaped towards supporting values, while the technological transformation can be designed according to the societal needs, not vice versa. This is especially important as ongoing societal developments in the fourth industrial revolution change the way value is created, exchanged and distributed. Further, technologies in Industry 5.0 must be regarded as part of systems that are actively designed towards empowering societal and ecological values, not technologies that determine societal developments. For example, the primary focus of technologies used should not be to replace the worker on the shop floor, but to support the workers' abilities and lead to safer and more satisfying working environments [14].

The technologies at the core of Industry 5.0 are largely congruent with Industry 4.0, while a stronger focus on human-centered technologies forms the basis for Industry 5.0. The term is further associated with the Japanese concept of Society 5.0, describing societal developments following or accompanying Industry 4.0 technologies. Industry 5.0 complements and extends the hallmark features of Industry 4.0. Its key dimension is the inclusion of a broader set of values, especially extending a human-centric towards a socio-centric perspective. This also increases the complexity and requirement for governance

Merging human and technological skills and strengths to the mutual benefit of industry and industry workers not technology replacing, but complementing humans. This allows safer, more satisfying and more ergonomic working environments, in which humans can use their creativity in problem-solving, adopt new roles, and enhance their skills. The core idea behind Industry 5.0 is to choose technologies based on an ethical rationale of how those support human values and needs, and not only based on what they can achieve from a purely technical or economic perspective. Technologies such as human-machine-interfaces, merging human brain capacities with Artificial Intelligence or collaboration with robots and machines are used to generate products and services. These products and services can be further customized to customer needs, reduce environmental impact and allow concept such as closed loops, energy self-sufficiency, emission-neutrality, or the Circular Economy.

While the development of the concept of Industry 5.0 started before Covid-19, it can be regarded as exemplary challenge that highlights the need to consider or reconsider concepts such as local generation of added value and reshoring, increased focus on resilience, optimization, return of stock and introducing the notion of sovereign capability in contrast to lean principles and productivity-driven rationales. Policy plays a crucial role in reshoring; it must create attractive framework conditions to stimulate companies to place their businesses [15].

Innovation and value creation must be oriented towards demands and needs, requiring organizational and societal change towards converging disciplines, mind-sets, leadership and decision making, and organizational silos, advancing concepts such as the shared economy. Further, trust in technologies, cybersecurity and data protection must be enhanced by encouraging customers to make informed decisions on the technologies and products they use. Moreover, individual trust in new technologies increases with the level of competence in maneuvering those, so substantial trainings and upskilling are needed across supply chains and, finally, developing the currently dominating techno-deterministic rationale into a human-deterministic.

2.5   Humans Closer to the Design Process of Manufacturing

As production systems will involve increasing levels of automation, informatics, robotics, sensors, mobile devices, etc., it's important to remember that human skills will remain essential for many tasks, making the marriage between humans and machines critical to success.

Human factors will therefore play an essential role in the future of manufacturing where people and technology are being integrated more closely and more intensively than ever before; it's essential that we fully understand how to best design and operationalize both human and technological functions

'Human Factors' (ergonomics) provides a scientific approach to human-centered design, applying physiological and psychological principles to optimize the balance of people's strengths and limitations.

It has a long history of making important contributions to the design of human-centered systems including manufacturing technologies and processes, although often as a limited part of engineering design.

The current challenge for industry is to include human factors with engineering and technology developments to optimize how workforces and shop floor environments are prepared for the transformational changes being brought about by augmented digitization and smart systems [16].

At our recent event held at Cranfield University to launch a new Manufacturing Sector Group, CIEHF members were joined by a number of engineers and industrialists to explore this topic together. Iain Wright, chair of Parliament's Business, Energy and Industrial Strategy Select Committee, opened the event by reminding us that Britain's manufacturing output still ranks as ninth in the world.

Human factors and ergonomics have a key role to play in reshoring manufacturing back to the UK. Wright explained: "While technology has the ability to increase productivity and reduce costs, it's human factors that will enable us to fully integrate our supply chains and enable differentiation."

2.5.1   Enablers of Industry 5.0

(a) **Exoskeleton** These are the devices and equipment human can wear on body to overcome fatigue and exhaust. Some exoskeletons may be driven by motor and actuators to reduce human efforts.

(b) **Block chain-enabled Fog computing** Blockchain technology provides extreme security feature. Information saves in blocks and blocks are connected to each other by unique id number (UID)

(c) **IOT Data interoperability** From Industry 4.0 we found that big data analysis helps in data analysis received from various IoT devices to achieve the mass customization faster

(d) **Intelligent automation** Robots that can directly interact to human by using safety features assured by human in them

(e) **Personalized delivery system** Drones are devices used in sudden inspection. These devices can also be used to transfer low weight material from one place to another place

(f) **Manufacturing traceability** Shop floor trackers are devices used to track the real-time production. They help to link the sales and production department along with buffer with efficient use of resources and reduced wastage

(g) **Mixed reality** Mixed reality is the outcome that is received when we integrate the physical world with the digital world.

It is the evolution that will involve the role of human, environment and computer interaction.

2.6   Challenges and Enablers (Socio-Econo-Techno Justification)

In general, the challenges and enablers discussed encompass a complex system across technological and organizational aspects, political and public factors, and the Triple Bottom Line of sustainability (economic, ecological and social dimensions). Therefore, the following categories of challenges and enablers represent a complex system and must be regarded as strongly interlinked and interrelated.

2.6.1   Social Dimension

☐ Technology acceptance and trust in technologies is crucial. Therefore, initiatives must highlight the support, not determination of humans, while maintaining the right for society to participate in the ideation and application of technologies and the respective purpose. In addition, new technologies such as Artificial Intelligence need to be understandable and transparent.

☐ Adapting technology to humans must coincide with training people how to use new technologies. Human-centricity should not be a one-way road. Otherwise, technologies will not unfold their full potential.

☐ Upcoming labor and skills shortages through demographic developments must be envisioned as well as changing requirements of skills across generations, while future skills required are not known yet. Retraining and lifelong learning concepts must be implemented, supported by policy makers.

☐ Current challenges such as youth unemployment, ageing society or gender and societal inequality must be integrated better to achieve a wide acceptance, forging a 37new deal between society and industry.

☐ Heterogeneity of society leads to disagreement on which values and needs should be prioritized. Different parts of society focus on different values and have distinct needs.

Age, gender, sex, cultural background and diversity aspects must be integrated.

☐ Full integration of customers across the entire value chains is required to inform them about social or environmental value created and include this in their choice and willingness to pay.

☐ Traditional Corporate Social Responsibility approaches are rather a marketing concept and implemented as showcases, but not implemented to full scale yet. Regulatory incentives must complement public opinions here to stimulate companies to drive Corporate Social Responsibility further, especially in SMEs.

☐ Socio-centricity describes a concept that extends the concept of human-centricity, complementing the needs of individuals to those of a society. The needs of individual employees must be integrated within the needs of an entire workforce of employees, employers, the society as a whole.

☐ In the long run, society cannot ignore planetary boundaries. Hence, socio-centricity must take into account the ecological perspective. Environmental targets such as CO_2 neutrality must not be forgotten and require research in the domain of e.g., nanotechnologies or smart materials. To develop approaches such as the Circular Economy further, corresponding business models, approaches such as CO_2 taxation and platform-based technologies for traceability, reusing and recycling are required. One option is a CO_2 footprint label on all products to make it possible to compare and get a personal or a company footprint.

☐ Environmental impact is hard to measure, social impact is even harder to quantify. Technological approaches must be found to measure and quantify environmental and social value generated.

2.6.2   Governmental and Political Dimension

☐ Social and governmental change often cannot keep up with the speed of technological change. This requires an 38agile government approach with issues such as responsiveness, collaboration, participation, experimentation, adaptability, and outcome-orientation based on intrinsic, extrinsic and market motivations.

☐ European industries and ecosystems should become more sovereign to facilitate compliance with European values, e.g. concerning resources, skills, raw materials, or energy. Regional value chains can help to ensure resilience and secure ecological and social values.

☐ There is a risk of protectionism of individual countries and industries in which there is a support to single industries rather than dynamic ecosystems, and certain technologies are selected not because of the values they contribute to, but for protectionist motives. Therefore, national or regional strategies must be widened regarding innovation policy design and implementation, for which RTOs and SMEs can take an active role.

☐ System-oriented innovation policies are required that focus on supporting ecosystems, not single technologies or sectors. In addition, policies must be evaluated based on their (possibly negative) interrelations with other policies or companies being overwhelmed with complexity. There is a clear need to strengthen capacity in the EU for system oriented innovation policy evaluation because it is the cornerstone for evidence-based and distributed intelligence in innovation policy-making.

☐ The complex ecosystem of technologies, industries and labor markets must be regarded from a holistic perspective and require a governance approach that integrates and balances the different requirements. For instance, social sciences and humanities (especially ethics) and the observation of labor markets must complement technological perspectives in order to support values.

☐ The economic perspective on productivity, that is dominant for many companies, must be balanced with a long-term, sustainability perspective on increasing productivity through generating ecological and social value. The need to be productive and competitive in a short to medium-term perspective in comparison to regions outside Europe, that might be mainly productivity-oriented and technology-driven, must not be forgotten in this context.

2.6.3   Interdisciplinarity

☐ An interdisciplinary approach is required related to engineering, technology, life sciences, and environmental and social sciences and humanities. One possible approach is to support interdisciplinarity of research from early on, e.g., the inclusion of social sciences in technological research. High complexity might otherwise have negative impacts on security, safety or acceptance and might slow down implementation, but fast actions are required.

☐ The interdisciplinarity and complexity requires a system approach, such as 40cyber-physical systems of systems. Such a System of Systems approach must model the interrelations at various scales of systems that dynamically interact. Such a modelling approach is challenging from various angles, as there is a wide variety of systems, scales and data involved as well as path dependencies arising from different industries and research disciplines.

☐ It is important to highlight that Industry 5.0 also includes sectors outside the manufacturing industry, such as life sciences, healthcare, agriculture, food or energy. Further, the inclusion of governments, consumers, and society is key to gain acceptance. Especially a biological transformation might lead to new values, but might not be associated to the term 'industry'.

☐ Artificial Intelligence needs to be included in research and design processes while integrating the requirements of different research disciplines in complex systems. In particular, the interrelations and cause-effect relations between multiple variables must be understood as dynamic networks that constantly change.

2.6.4   Economic Dimension

☐ Business models need to be developed that make ecological and social value approachable, e.g. through benefitting from social and environmental value created as the customer is willing to pay for it, or through legislation, such as CO_2 certificates. Companies are required to make money. Such business models can be complemented with digital platform approaches and ecosystems that allow the integration of multiple stakeholders, such as companies, public institutions, and customers.

☐ Productivity is still required to maintain competitiveness. Economic and social value generated is becoming an increased competitiveness factor. Therefore, the economic dimension must not be neglected and economic targets should not be left out of the concept for companies. Otherwise, only marketing-driven showcases might be implemented while many firms could be discouraged from the concept.

☐ Large investments are required that follow a new rationale in contrast to purely economic considerations. Solutions must be found to, for instance, guide private equity funds towards generating social and environmental added value.

☐ In this context, public-private-partnerships may be a suitable instrument to institutionalise the dialogue between business and policy in specific areas.

2.6.5   Scalability

☐ Many Industry 4.0 technologies are not fully implemented yet, especially in SMEs or across entire value chains. Further, innovation management and R&D investments are underdeveloped and not standardised, particularly in SMEs or traditional industry sectors. The policies should also support broad implementation in entire industries or in SMEs, not only High-Tech for a small proportion of firms. For this, a wider integration of industry associations and representatives from industry is required for the further design of the concept. Skills in leadership and management alongside workers skills must be further developed.

☐ Entire industries and ecosystems must be addressed with a systematic approach to new technologies. Entire ecosystems must be understood and supported accordingly, including leading players, followers, SMEs, RTOs and universities as well as legislation and policy makers. A full-scale implementation at a lower technology level can bring more benefits in comparison to single pillars that can be leading examples, but must be followed by broad implementation.

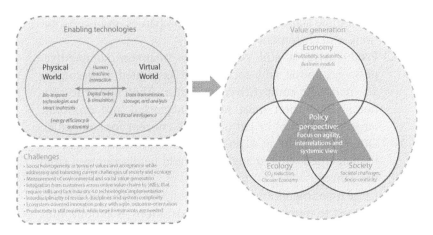

Figure 2.1 Framework depicting Industry 5.0.

When technologies that stem from life sciences are combined with engineering and information technology disciplines, this will require a more systematic innovation approach that integrates different perspectives and takes a systemic view on entire ecosystems. Further, the systems generated will be highly complex, interrelated, and interdependent and will have to cope with inhomogeneous data sets. Economic targets such as productivity and competitiveness must not be neglected but set within agreed ecological and social values. This can be achieved through business models which value ecological and social value creation or incentives from legislation.

Apart from nomenclature and challenges arising from complexity or technological considerations, society and the bulk of the industrial landscape must be integrated in such a concept. Therefore, customers and entire supply chains, up to SMEs must be better integrated to ensure a broadscale implementation and value generation towards prosperity [17, 18].

Figure 2.1 highlights the main features of goals, technological enablers, and challenges associated with the concept of Industry 5.0.

2.7   Concluding Remarks

Industry 5.0 is a new production model where the focus lies on the interaction between humans and machines. The previous tier, Industry 4.0, emerged with the arrival of automation technologies, IoT and the smart factory. Industry 5.0 takes the next step, which involves leveraging the collaboration between increasingly powerful and accurate machinery and the unique creative potential of the human being. The phase that came before Industry 5.0 has seen the emergence of digital industry: advancements such as the Industrial Internet of Things or the combination of Artificial Intelligence and Big Data have generated a new type of technology that can offer companies a data-based knowledge. This has, in turn, translated into processes such as Operating Intelligence and Business Intelligence, which generate models that apply technology with aims on making increasingly accurate and less uncertain decisions.

However, during this phase in Industry 4.0, the goal has been to minimize human involvement and prioritize **process automation**. To a certain extent, humans have been set in a position where they competed against machines, casting the former aside in a myriad of scenarios. In the case of **Industry 5.0**, this trend is reversed: the goal is to strike a **balance** whereby the machine-human interaction can offer the highest benefits. The changes set in motion by Industry 5.0 are already irreversible. This process offers companies the abilities of **increasingly powerful** machines in combination with better-trained experts to foster an effective, sustainable and safe production. Industry 5.0 is not a fad, but rather a new way of understanding manufacturing that has productive, economic and commercial consequences.

Therefore, companies that do not **tailor their production** to the factory 5.0 model will soon become obsolete, being unable to benefit from the competitive advantages that it has to offer. Not only that: the **rate of technological acceleration** is increasingly faster and shows that the emergence of new paradigms never stops. For this reason, adjusting the processes of each company and transforming them into the concept of digital industry will be vital in guaranteeing that an organization remains competitive.

References

1. Chen, S.H., Jakeman, A.J., Norton, J.P., Artificial intelligence techniques: An introduction to their use for modelling environmental systems. *Math. Comput. Simul.*, 78, 379–400, 2008.
2. Jordan, M.I. and Mitchell, T.M., Machine learning: Trends, perspectives, and prospects. *Science*, 349, 255–260, 2015.
3. Kavousi-Fard, A., Khosravi, A., Nahavandi, S., A new fuzzy-based combined prediction interval for wind power forecasting. *IEEE Trans. Power Syst.*, 31, 18–26, 2016.

4. Khosravi, A., Nahavandi, S., Creighton, D., Prediction intervals for short-term wind farm power generation forecasts. *IEEE Trans. Sustain. Energy*, 4, 602–610, 2013.

5. Khosravi, A., Nahavandi, S., Creighton, D., Prediction interval construction and optimization for adaptive neurofuzzy inference systems. *IEEE Trans. Fuzzy Syst.*, 19, 983–988, 2011.

6. Kotsiantis, S.B., Zaharakis, I., Pintelas, P., Supervised machine learning: A review of classification techniques. *Emerg. Artif. Intell. Appl. Comput. Eng.*, 160, 3–24, 2007.

7. Nayak, N.G., Dürr, F., Rothermel, K., Software-defined environment for reconfigurable manufacturing systems, in: *Proceedings of the 2015 5th International Conference on the Internet of Things (IOT)*, Seoul, Korea, 26–28 October 2015, pp. 122–129.

8. Nahavandi, S., Robot-based motion simulators using washout filtering: Dynamic, immersive land, air, and sea vehicle training, vehicle virtual proto-typing, and testing. *IEEE Syst. Man Cybern. Mag.*, 2, 6–10, 2016.

9. Nguyen, T., Khosravi, A., Creighton, D., Nahavandi, S., Spike sorting using locality preserving projection with gap statistics and landmark-based spectral clustering. *J. Neurosci. Methods*, 238, 43–53, 2014.

10. Nguyen, T., Khosravi, A., Creighton, D., Nahavandi, S., Hidden Markov models for cancer classification using gene expression profiles. *Inf. Sci.*, 316, 293–307, 2015.

11. Nguyen, T. and Nahavandi, S., Modified AHP for gene selection and cancer classification using type-2 fuzzy logic. *IEEE Trans. Fuzzy Syst.*, 24, 273–287, 2016.

12. Papadimitriou, F., Artificial intelligence in modelling the complexity of med-iterranean landscape transformations. *Comput. Electron. Agric.*, 81, 87–96, 2012.

13. Singh, C.D. and Kaur, H., Application of modified fuzzy TOPSIS in optimiz-ing competency and strategic success. *Int. J. Manage. Concepts Philosophy*, 14, 1, 20–42, 2021.

14. Singh, C.D., MCDM based modelling for sustainable green development through modern production techniques. *Int. J. Compet.*, 2, 1, 62–90, 2021.

15. Sulema, Y., Asampl: Programming language for mulsemedia data pro-cessing based on algebraic system of aggregates, in: *Interactive Mobile Communication, Technologies and Learning*, pp. 431–442, Springer, Berlin, Germany, 2017.

16. Wang, J., She, M., Nahavandi, S., Kouzani, A., Human identification from ECG signals via sparse representation of local segments. *IEEE Signal Process. Lett.*, 20, 937–940, 2013.

17. Yetilmezsoy, K., Ozkaya, B., Cakmakci, M., Artificial intelligence-based prediction models for environmental engineering. *Neural Netw. World*, 21, 193–218, 2011.
18. Zhou, H., Kong, H., Wei, L., Creighton, D., Nahavandi, S., Efficient road detection and tracking for unmanned aerial vehicle. *IEEE Trans. Intell. Transp. Syst.*, 16, 297–309, 2015.

3

Machine Learning – A Survey

Navdeep Singh* and Aanchal Goyal

Department of Computer Science and Engineering, Punjabi University, Patiala, Punjab, India

Abstract

Machine learning has grown rapidly in the last decade and holds importance not only in the technological industry but even in factories and manufacturing units. It is a notion that has been around for quite long period of time. The notion of automating the application of sophisticated mathematical computations to massive data, on the other hand, has only been around for a few years, but it is getting momentum today. Machine learning is most often utilized by organizations when they implement artificial intelligence systems nowadays, and the phrases machine learning and artificial intelligence are commonly used interchangeably. Machine learning is an artificial intelligence area that allows machines to understand without being explicitly programmed. The growing interest in machine learning is due to several factors such as cheaper computational processing power, high availability of the data, and affordable storage for the data. All of this ensures that models that can evaluate larger, more complicated data and offer faster, quite accurate answers may be created rapidly and automatically, even on a massive scale. An organization's chances of recognizing profitable possibilities or avoiding unforeseen hazards, improve when detailed models are built. The importance can be linked to the accurate predictions without human intervention which can ultimately help take smart actions and make better decisions in the real world. The usefulness of machine learning technology has been acknowledged by most businesses that deal with big volumes of data. Organizations might operate more effectively or gain competitive edge over rivals by harvesting insights from this data frequently in real time. Machine learning is significant because it allows businesses to see insights about customer behavior and company operating patterns while also assisting in the creation of new goods. Every market is evolving, or will evolve, and professionals must

---

Corresponding author: navdeepsony@gmail.com

Chandan Deep Singh and Harleen Kaur (eds.) *Factories of the Future: Technological Advancements in the Manufacturing Industry*, (47–82) © 2023 Scrivener Publishing LLC

comprehend the fundamental concepts, promise, and constraints of machine learning. In this chapter, various types of machine learning and associated techniques have been discussed. The chapter discusses not only supervised but also unsupervised techniques in a comprehensive manner so as to give the reader a solid understanding of the fundamental concepts. Different types of clustering techniques have also been considered along with their advantages and disadvantages. Feature reduction techniques to reduce the dimensionality of the data and various distance measures are extensively covered in the chapter. The chapter provides an insightful understanding to the researchers about numerous applications of machine learning and future scope.

Keywords: Artificial Intelligence, machine learning, automation, data analysis, prediction, feature extraction

3.1 Introduction

Machine learning is the field of AI that makes computers get into a self-learning mode without being explicitly programmed. It has been around from quite a long time now, evolving continuously along with the increasing demand and importance. In this chapter, various types of machine learning techniques and applications are discussed in the first section with the help of real-world examples. For machine learning, a huge amount of data is available which isn't purely clean and accurate. The data needs to be processed before being used. Some of the data has excessive features than what is actually required. With the excessive features, various problems arise such as overfitting, curse of dimensionality etc., which needs to be dealt with dimensionality reduction. Machine learning analysis is to evaluate and try the machine modelling in the various expressions, which smoothens the evolution of the intelligent systems. In this chapter, clustering-based machine learning algorithms are elaborated in detail. The data is portioned into groups based on similarities in data and these groups are labelled. Clustering is a technique that offers unsupervised learning often called as a data analysis technique used for discovering interesting patterns in the data, for instance spam filtering of emails. Clustering analysis is broadly employed in various applications such as market segmentation, anomaly detection and statistical data analysis. Clustering methods are broadly split into Hard clustering in which data points are related to only one group and Soft clustering in which the data points are related to even other groups. The types of clustering discussed in this paper include the following:

- Iterative distance-based clustering
- Density-Based Clustering
- Hierarchical Clustering

In the last section of the chapter, future applications of machine learning are discussed based upon manufacturing units and industries. It covers how machine learning is applied in various fields and what enhancements it requires to solve various problems and boost the production not only in the technical field but other fields as well.

In section 3.1, a detailed discussion on machine learning is provided followed by section 3.2 which contains a detailed report on dimensionality reduction. Section 3.3 covers various distance metrics and their types. Sections 3.4-3.7 introduces the clustering algorithms, including introduction to clustering, iterative distance-based algorithm, density-based algorithm and hierarchical algorithm. Section 3.8 covers the use of machine learning in future industries and provides a direction for future work.

3.2   Machine Learning

A huge chunk of data is being collected nowadays in various forms and from various devices such as mobile phones, laptops, GPS and other devices. With the availability of huge data, we can generate many machine learning models to accomplish various tasks such as prediction, forecasting etc. Machine learning has been around us for quite some time now. It has been evolving continuously along with increasing demand and importance. It helps us to understand the structure of data and build models accordingly that can be understood by humans. The machine learning algorithms can make predictions through pattern recognition. It provides a meticulous way of solving problems through identification, classification or prediction. Machine learning is the sub-domain of artificial intelligence (AI) in which the systems can make intelligent decisions without being explicitly programmed with the help of experience and feedback [1].

It is divided into three parts:

- Unsupervised Learning
- Supervised Learning
- Reinforcement Learning

3.2.1   Unsupervised Machine Learning

Unsupervised learning is a kind of machine learning where the training of a model is done using unclassified and unlabeled information and it allows the algorithm to act upon information without guidance [2]. Here the task of the machine is grouping of the information that isn't sorted as per patterns, differences and similarities with none prior training of data. The major aim of unsupervised learning is finding the fundamental structure of the dataset, then grouping the data in accordance to its resemblance and representing the dataset in an abbreviated layout. No supervision in terms of input and output is provided to the model and therefore, the machine is restricted to find the hidden patterns and structures in unlabeled data solely on its own.

Various features of unsupervised learning are:

- It helps in locating useful intuitions from the data
- It is similar to the way as humans learn from experiences and practice, which makes it closer to AI.
- The unlabeled and uncategorized data is fed into it.
- In real-world, the output data is not always available along with the input data and the system or model has to find the patterns automatically by itself and this is where unsupervised learning plays a significant role.

Its working can be understood via simple example:

In this Figure 3.1, there is unlabeled data input, that means it is not classified and the corresponding output is not labelled with it. Unsupervised

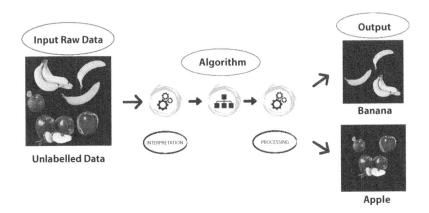

Figure 3.1 Unsupervised learning [1].

learning models are fed with this data in order to train it. Primarily, the raw data is interpreted by the model in order to find the concealed patterns among the information and then a suitable algorithm is applied to the same such as, decision trees, k-means clustering, etc. After the application of a suitable algorithm, data objects are divided into categories as per resemblance and distinctions among the objects.

3.2.2 Variety of Unsupervised Learning

Unsupervised algorithms are classified into two types:

i. Clustering: The technique of categorizing the objects into clusters in such a way that items with the most correlations are put into a grouped cluster and ones with lesser resemblance are segregated into another group. Clustering looks for the similarities between the data objects, which acts as a base to categorize the data objects.

ii. Association: It is a method of unsupervised learning used to decree the analogy between variables in the large database. It figures out which objects are occurring together in the dataset. Marketing strategy can be made more effective by association rules. Market Basket Analysis is one of its known examples.

Various unsupervised learning algorithms are as follows:

- K-means clustering
- KNN (k-nearest neighbors)
- Hierarchical clustering
- Independent Element Analysis
- A priori method
- Singular value decomposition

Advantages of unsupervised learning:

- Providing labels for the dataset is too much of a task. This issue can be resolved by unsupervised learning and the classification of datasets can be done without labels.
- The algorithm adds labels to the output of the dataset by learning all by itself.
- Patterns in a dataset can be found using it which cannot be done by any other normal method.

- In it, accomplishing dimensionality reduction is quite easy.
- By using probabilistic methods, degree of similarity can be found for the given dataset.
- It is much like the human brain in some way as it functions similarly to it by learning slowly and then processing the result.

Disadvantages of unsupervised learning:

- The accuracy of the result might not be that good as there is no input data to train.
- No prior knowledge about the dataset is given to the model.
- A lot of time is taken by the algorithm to learn, as all the possibilities are calculated by it.
- With the increase in features, complexity increases as well.

3.2.3   Supervised Machine Learning

In it, both the input in addition to output data is available to the model using which the model is trained and as a result the model makes very accurate predictions. Consequently, the machine is supplied new set of data with the purpose that the algorithm examines the training data and provides accurate output from labelled data. The main objective of supervised machine learning technique is to discover a mapping basis to map the input data (x) to the output data (y). Mathematically, supervised learning can be shown as a linear function, i.e., $y = f(x)$. In the real world, it is used in various applications such as Risk Assessment, Image Prediction, Spam Filtering, etc. Formally, supervised learning can be defined as an algorithm that learns from labelled training data, which helps in the prediction of outcomes for unforeseen data which belongs to the same group as in the training set [3].

Various features of unsupervised learning are:

- It provides a direct path for turning data into real and practicable insights.
- By using data, it allows organizations to comprehend and prevent unfavorable results.
- It boosts the desired outcome by prediction through the provided data.
- After training, the machine gains experience and can be used to perform prediction on the similar unseen data.

- The performance of the algorithm is optimized by experience.
- Supervised learning can take care of real-world computations.

In the supervised study, models are instructed on a labelled dataset, in which the model learns about every type of data. As soon as the training is completed, the model is evaluated on the validation data as shown in Figure 3.2. In this example, there is a dataset with three dissimilar types of emoticons that includes happy, sad, and angry. The initial step is to make the model learn each emoticon which is done via image prediction.

- If the emoji is red, with sharp eyes and upside-down smile, it is labelled as angry.
- If an emoji has an upside-down smile and yellow color, it is labelled as sad.
- If an emoji has an upward smile and yellow color, it is labelled as happy.

Once the model is trained, it is applied on the test set where the task of the model is to identify the type of emoticon. As the machine is already trained on different kinds of emoticons using the features such as color and shape, it will be able to correctly classify the emoticon as happy, angry or sad.

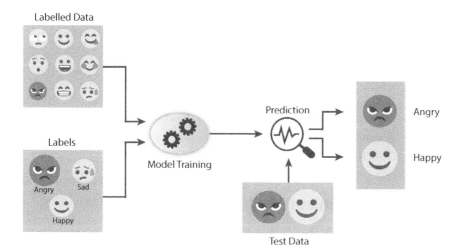

Figure 3.2 Supervised learning [3].

3.2.4 Categories of Supervised Learning

Supervised learning can further be divided into two types:

i. Regression: It is used when there is relation between the dependent and independent variables of the dataset, considering that the variables are continuous. Some of the examples of the use of regression are forecasting of weather, market trends, etc. Some of the famous regression techniques are:
    - Linear Regression
    - Regression Trees
    - Non-Linear Regression
    - Bayesian Linear Regression
    - Polynomial Regression

ii. Classification: It is used when the yield of the data is categorical, and that means there are only two possible categories such as Male – Female, Yes - No, 0 – 1, even – odd, etc. Spam filtering is an example of classification. Some of the classification algorithms are:
    - Random Forest
    - Decision Tree
    - Logistic Regression
    - Support Vector Machines

Advantages of Supervised Learning:
- It is easy to understand.
- The number of groups is already known in the given data.
- It is very helpful in solving real-world computational problems.
- These models learn from previous experiences and therefore help improve the accuracy of prediction.

Limitations of Supervised Learning:
- These are not suitable for handling complex tasks.
- Training requires lot of time for computations
- We need enough knowledge about the classes of objects.
- It cannot create labels of its own.
- The new data must be from the given classes only.

3.3 Reinforcement Machine Learning

Reinforcement Learning is a sub-field of machine learning which deals with how intelligent agents act in the given environment with the aim to maximize the idea of cumulative reward [4]. Its main purpose is to detect a balance between exploitation of current knowledge and exploration of uncharted territory. Both the reinforcement and supervised learning uses mapping among inputs and outputs but reinforcement differs from supervised learning by not requiring labelled dataset and sub-optimal actions to be explicitly corrected. Rewards and punishment are used as signs by reinforcement learning for both positive as well as negative practices. In comparison with unsupervised learning, it differs in respect to goals, the objective of unsupervised learning is to discover resemblance and dissimilarities among data points, whereas in reinforcement learning the objective is to build an appropriate model for action that would enhance the total cumulative reward of the agent.

For the formulation of a reinforcement learning problem, a few elements are required namely:

Environment: An agent operates in a physical world known as the environment.

State: The current stage of the agent is known as a state.

Reward: Reward is the positive or negative feedback received by the agent for its actions [5].

Value: It is the future benefit that an agent would acquire for its action depending upon the action. Policy: It is a method for mapping the action states of an agent.

The best example of reinforcement learning is the game known as "Pacman" as shown in Figure 3.3. The aim of the agent in the game is to eat maximum food available without getting caught by the ghostly creature.

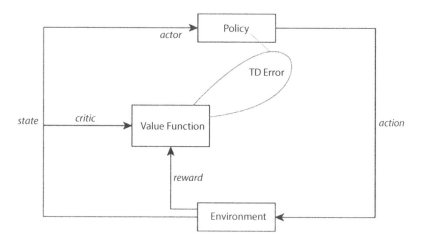

Figure 3.3 Actor-critic architecture [6].

Agent is provided with a grid like interactive environment, where the agent, ghost and food coexist. Agent moves around the grid to eat up maximum food for which it is rewarded and when it is caught by the ghost and killed, it serves as a punishment for the agent. The various locations of the agent in the environment are its states and the cumulative rewards serve the agent for winning the game.

To create an optimal strategy, the agent needs to explore new states while keeping the reward in view, which needs to be maximized. This particular concept is known as exploration Vs exploitation trade off [4]. Some of the frequently used reinforcement learning techniques are Q-Learning [4] and SARSA. These vary in respect of their exploration policies while their manipulation policies are similar. Here Q-learning stands for quality learning. It is an off-strategy reinforcement learning algorithm in which the representative seeks the best effort from outside of current policy. SARSA is said to be an on-policy procedure which learns from its current effort obtained from its current policy. The procedures lack in generality as they do not possess the ability to estimate values from unseen states but the main benefit associated with them is that they are really simple to implement. This inability to estimate values can be tamed by modern algorithms such as Deep Q-Networks that utilize Neural Networks to determine Q-values that can take care of discrete and low proportion action spaces. A Deep Deterministic Policy Gradient is a model-free, no policy, critic method that handles these issues by studying strategies in high dimensional and continuous spaces.

3.3.1   Applications of Reinforcement Learning

As per the requirement enough data for Reinforcement Learning, is mostly applicable in the domain areas where reproducible data is easily accessible such as gameplay and robotics, etc. [4].

- Its biggest application is developing AI for playing the games on computers. The very first program to conquer a world champion was AlphaGo Zero which is an ancient Chinese game of Go. In it the AlphaGo Zero was able to teach itself the game of GO after training on it for a period of months and after that it defeated the world champion of the game. Other examples include ATARI games, Zero-Sum [7], Prisoner's Dilemma Backgammon, etc.
- It is used in Industrial automation and robotics to enable robots to create an effective control system for themselves,

have the ability of grasping various objects, learning through its experience and behavior. Robotic handling with asynchronous policy updates is one such nice example of it. It was even used by Google AI to build seven real-world robots.

- Other applications include the field of marketing and advertising, news recommendation, text summarization engines, understanding the best treatment policies in healthcare and financial services agents etc.

3.3.2   Dimensionality Reduction

Dimensionality represents the number of input features or variables in a dataset. Dimensionality reduction is a technique that is used to cut down on the number of features present in the dataset. It is a method for reducing the complexity of a model and prevent overfitting. When the data is gathered, there are a lot of features which are not actually required or are not necessary for analysis and excessive features can cause the problem of overfitting of data. Features are the basic factors on which segregation is enacted. With the large number of features, visualization and working on the dataset becomes harder. Sometimes, most of the features are correlated and hence can be reduced. Therefore, we use a process referred to as feature reduction which is widely known as dimensionality reduction likewise to scale back the number of features in a resource, by obtaining a set of principal variables without losing the important information. By the reduction of features, it means to reduce the number of variables for making computational work easier, efficient and faster as shown in Figure 3.4. It leads to the need of fewer resources to complete computational tasks. As, less computation time, less storage capacity will be required which means more work can be done. It reduces multicollinearity resulting in an improved machine learning model.

Dimensionality Reduction has two components:

- Feature Selection: In it, a subset is found of the initial set of characteristics, to induce a subset which is compact and used to map the problem. It generally consists of Filter, Wrapper and Embedding.
- Feature Extraction: In it the dimensionality of that data is reduced to the lower dimension space by extracting various features. Lower dimension implies a state which has very less number of dimensions [9].

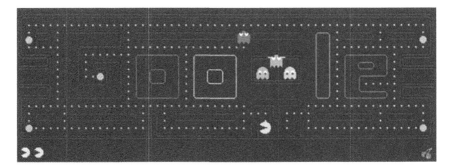

Figure 3.4 Game of PacMan.

3.4   Importance of Dimensionality Reduction in Machine Learning

This can very well be understood via the problem of classification that is based on both rainfall and humidity which can be reduced into a single feature as both of them have a high degree of correlation. Email classification problem is another such problem, in which there is a necessity to determine whether or not a message is spam. All the features of an email are checked including generic title, template used, the content of the mail etc. Sometimes features may overlap and therefore to avoid overlapping of features, the number of features needs to be reduced. Visualization of a 3D classification problem is hard, whereas a 2D and 1D problem can simply be assigned to a 2D dimensional area and a simple line respectively. It is illustrated in Figure 3.5, where a 3D data has been broken into two lower dimensional spaces which can further be reduced if more correlating features are found [9].

3.4.1   Methods of Dimensionality Reduction

Different methods that are used to reduce dimensionality are as following:

- Principal Component Analysis
- Linear Discriminant Analysis
- Generalized Discriminant Analysis

Depending upon the usage of method, dimensionality reduction can be linear or nonlinear.

3.4.1.1   Principal Component Analysis (PCA)

Principal Component Analysis is also referred to as the Prime linear method. This mechanism was developed by Karl Pearson. It operates on

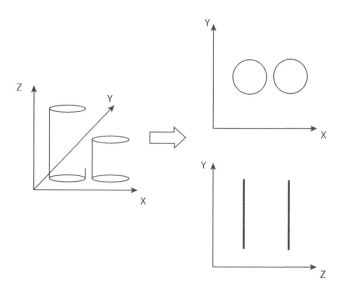

Figure 3.5 Components of dimensionality reduction [8].

the situation that the information in higher dimension can be projected into a lower dimension with a condition that the data in lower dimension must have maximum variance. It seeks out the optimal linear combinations of the original variables in order to maximize the variation or spread across the new variable.

Following steps are involved in the calculation of principal components:

- The covariance matrix is constructed for the data.
- Computation of eigenvectors for the constructed matrix.
- The Eigenvectors which have largest corresponding eigenvalues are used to create the principal components.

As can be observed from Figure 3.6, there are fewer eigenvectors, and there is the possibility of losing some data in the process but what matters is that the variances are retained by the remaining Eigenvectors.

3.4.1.2   Linear Discriminant Analysis (LDA)

Linear Discriminant Analysis is another method for dimensionality reduction which is very similar to Principal Component Analysis with an added advantage that it achieves maximum separation for classes and minimum

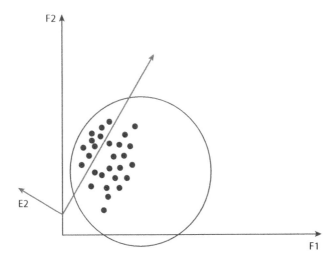

Figure 3.6 Principal Component Analysis [8].

separation within each class which ensures better features reduction. The multiclass objects or events are categorized by the combinations. The resulting combination is used for dimensionality reduction. It attempts to model the difference between the classes of information and uses matrix factorization at the core of the technique. It includes statistical properties of the data, computed for each class where the goal of LDA as a supervised algorithm is to discover a characteristic subspace that improves class separation.

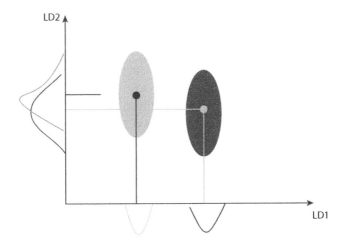

Figure 3.7 LDA: maximizing the component axes for class-separation [10].

It is used for the projection of features in higher dimensional space into lower dimensional space. For a single input variable (x), this is the mean and variance for each class and for multiple variables, the means and covariance matrix are formulated. After the estimation of these statistical properties, prediction is made after substituting in LDA equation.

In the Figure 3.7, X-axis i.e., LD1 represents the reduced dimensionality and y-axis i.e., LD2 represents a new component in the reduced dimensionality. Thus, LDA separates the two normally distributed classes very well.

3.4.1.3   *Generalized Discriminant Analysis (GDA)*

Data contains a large amount of ineffective information; therefore, we need to remove worthless information from the high dimension. With the help of GDA not only the number of input features is reduced but the accuracy of classification is also increased and reduction in the time period of testing and training is done by selecting most discriminant features. Generalized Discriminant Analysis is a non-linear discriminant analysis which is based upon kernel function which maximizes the ratio of between-class to within-class dispersion to project the data matrix from a high-dimensional region into a low-dimensional region. The mapped data is transformed to high dimensional space where different classes are linearly separable. The LDA scheme is then applied to the mapped data, and ultimately those vectors are calculated which best discriminates the data.

3.5   Distance Measures

Distance measures play a very crucial role in machine learning. It serves as a well-built base for several machine learning algorithms such as Knn and k-means etc. Performance of a machine learning model can be boosted with an effective distance metric and it does not matter if the problem is of supervised or unsupervised learning. Distance measure is a metric that evaluates and compares two objects in a problem domain. Depending upon the category of data, distinct appropriate distance metrics are chosen which includes some basic requirements such as the distance is non-negative, symmetric, identity of indiscernible and triangle inequality [11]. By having the knowledge of which metric is to be used, can make one go from poor classifier to a pretty perfect model [12]. Therefore, it is necessary to understand the intuitions behind various distance metrics. Some of the distance metrics are mentioned below:

- **Euclidean Distance**

It is the most common distance measure. It is the representation of the shortest distance between the two vectors [13]. The distance can be calculated from Cartesian coordinates of the two points via Pythagoras theorem and is shown below in Figure 3.8 and the formula as the square root of the sum of the squared dispersions between the two points.

$$D(x, y) = \sqrt{\sum_{i=1}^{n} (x_i - y_i)^2}$$

where, n is the number of dimensions; x_i, y_i are the data points.

Despite being a common distance measure method, the distance between the points might be skewed depending upon the units of the features. Therefore, normalization of the data is necessary prior to its use. It is not useful for the high dimensional data; it only works well for lower dimensional data and is useful when the magnitude of a vector is necessary to be measured. It is quite simple to implement.

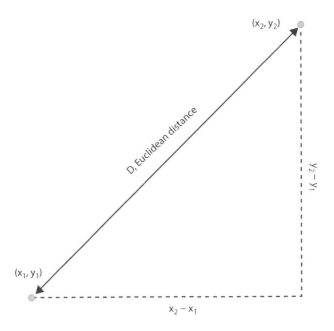

Figure 3.8 Euclidean distance [13].

- **Manhattan Distance**

Manhattan distance is nothing but the sum of the absolute dispersions between the points across various dimensions. It can be calculated only if the movements of the vector are perpendicular [14] just like the vectors that describe the uniform grid of chess, no diagonal movement is involved in the calculation of this distance. City block distance and Taxicab distance are its other names [13].

The generalized formula for Manhattan distance for an n-dimensional space is

$$D_m = \sum_{i=1}^{n} |p_i - q_i|$$

where, n is the number of dimensions; p_i, q_i are the data points.

It works fine in the higher dimensional data but is less intuitive while dealing with high dimensions. It should be used for the data having binary or discrete attributes as it takes a path in the account which could be taken within the parameters of the attribute realistically.

For travelling from point P to Q the path is not in the straight line. Therefore, the Manhattan distance metric can be used where the blue line gives the Manhattan distance. This distance is calculated as shown in the red lined path in Figure 3.9.

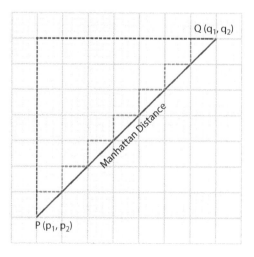

Figure 3.9 Manhattan distance [14].

- **Minkowski Distance**

It is much more intricate measure and is a generalized form of both Euclidean and Manhattan distance measures used in normed vector spaces i.e., n-dimensional real space which signifies its usage for the spaces where the distance can be represented as a vector having a certain length as shown in Figure 3.10. By adding the order parameter, p, different distance measures can be computed. There are three basic requirements for this;

- Triangle Inequality: It states that the shortest distance between two points is a straight line.
- Zero Vector: It has zero length unlike other vectors having positive length just like in the case of displacement when the initial and final positions are the same.
- Scalar Factor: Positive number multiplied with the vector which increases the distance while keeping the direction same.

Its formula is:

$$D(x,y) = \left(\sum_{i}^{n} |x_i - y_i|^p \right)^{1/p}$$

where, n is the number of dimensions; x_i, y_i are the data points, p is order parameter. The common values of p are: 1 – Manhattan distance, 2 – Euclidean distance, ∞ – Chebyshev distance [12].

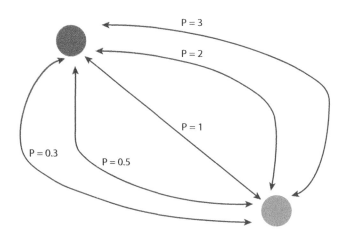

Figure 3.10 Minkowski distance [12].

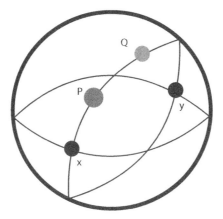

Figure 3.11 Haversine distance [12].

- **Haversine Distance**

It is the type of distance metric that measures the number of mismatches between the two vectors [15]. It is the distance between two points on a sphere, with their longitudes and latitudes as shown in Figure 3.11. No straight-line formation takes place as the points are assumed to be on a sphere.

Haversine distance is calculated as shown below;

$$D = 2rarcsin\left(\sqrt{sin^2\left(\frac{\varphi_2 - \varphi_1}{2} \right) + cos\,cos(\varphi_1)\,cos\,cos(\varphi_2)sin^2\left(\frac{\lambda_2 - \lambda_1}{2} \right)} \right)$$

Here, φ_1, φ_2 is the latitude of the two points λ_2, λ_2 is the longitude of the two points.

The biggest disadvantage associated with it is that the points are assumed to be lying on a sphere as in some cases the figure is not properly round which can create issues during calculations. It is mostly used during navigation such as calculating the distance for flying between two places.

3.6   Clustering

Clustering is a method to segregate the populated data points in such a way that the data points in a group have the most similar properties but dissimilar to other groups. In short, it is a way of clustering data according

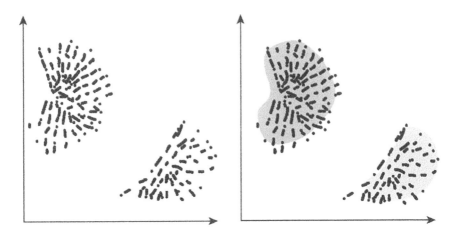

Figure 3.12 Clustering the data points on the basis of density.

to the similarities of data into different clusters. It is illustrated in Figure 3.12 given below in which the data is clustered based on its density.

Based on the type of clustering, it can be classified into two types:

- Hard Clustering: The kind of clustering in which the data can either belong to a group or cannot, data is exclusively grouped i.e. a data point belongs to a definite cluster and cannot belong to another cluster [16]. For example, in the clustering of dogs and cats, either the given data point is a cat or a dog but it cannot be both.
- Soft clustering: In which the data point is put into a cluster based upon the likelihood or probability of the data point to be in that cluster. Fuzzy sets [17] are used for clustering in such a way that each data point may belong to more than one cluster with different intensity [16]. Soft clustering is more frequently used than hard clustering and is also called overlapping clustering.

Clustering holds a lot of importance in order to determine intrinsic grouping of the unlabeled dataset. One can choose among various methods having their own set of rules, which are stated below:

- Centroid Models: It is a repetition based clustering [18] where the similarities between the data points are checked by the closeness of the data point to the centroid of the clusters. In it, the object is partitioned into k clusters where the number of clusters is known before-hand, therefore one requires prior knowledge of the dataset. One of the well-known algorithms using centroid is K-means clustering.

- Density Models: It is based upon the notion that searches for the data spaces for varied density of data points. In it the clusters are formed based upon the data points present in the dense region, where the dense regions have similarities when compared with the less dense regions in the dataset. This method has good accuracy and ability for merging two clusters but it is hampered by the problem of overfitting which may arise in it. The most commonly used density-based algorithm is DBSCAN [18].

- Distribution Models: In it, the clustering is based on the notion of distributing data points on the probability of it being in that group. Such models generally suffer from overfitting. The commonly used example is Expectation – maximization algorithm that uses multivariate normal distributions.

- Hierarchical Based Model: In it the clusters form a tree-like structure known as dendrogram [18] which is based on hierarchy and each new cluster formed is based upon the parent cluster. It is further divided into two categories, Agglomerative and Divisive clustering, which are bottom up and top-down approaches respectively [18] which are discussed in detail further in this chapter. Examples for the same are Agglomerative hierarchical algorithms and CURE (Clustering using Representatives).

- Grid-Based Model: In it a grid is made up of a finite number of cells like structure formulated from the dataset. The operations carried out on these clusters are fast and are independent of the number of data objects. Wave cluster, CLIQUE, STING are its examples.

- Fuzzy Clustering: It is a soft method where the data points may be a part of many clusters. Here, each dataset has a series of participation coefficients that are based on the amount of cluster affiliation. Fuzzy k-means technique is one of the known examples.

3.6.1   Algorithms in Clustering

Clustering algorithms can be divided based upon their methods. Algorithms depend upon the type of dataset used, such as in some datasets, the requirement is to determine the shortest distance between measurements of a dataset and in some there is a need to guess the multitude of groups. Some of the popular algorithms are described below:

- K-Means: It is based upon the partitioning method; in it clusters are classified by separating the dataset into distinct equal-variance clusters. Here it is necessary to specify the number of clusters in the algorithm. It requires lesser number of computations and is faster with a linear complexity of $O(n)$.
- Affinity Propagation: It is not like other methods as in it the specification of number of clusters is not required. In it a message is sent between a pair of data points by each data point until convergence. The time complexity of this algorithm is not linear i.e., $O(n^2)$, which can be seen as a drawback .
- Agglomerative Hierarchical Algorithm: It is based upon Hierarchical based clustering using bottom-up approach. In it, each data item is considered a separate cluster and afterwards merged consecutively.
- Exception-Maximization Clustering via GMM: It is used as a substitute to approach for the k-means method for the places where k-means fails. In it, the data points are supposed to be Gaussian dispersed, which means it is a probability distribution that is symmetric about the mean.
- Mean-Shift Algorithm: It is a centroid based approach in which the dense region with smooth density of data points is opted, where the centroid is located within the given region.
- DBSCAN Algorithm: It is a density-based approach which is the same as the mean shift but with many benefits. In it the areas with higher density are divided from regions with low density and as a result of this variation, the clusters can be seen in a variety of shapes.

3.6.2   Applications of Clustering

1. Biology: Used to classify various specimens of flora and fauna via image recognition technique.
2. Customer Segmentation: It is employed for research into the market where the customers are clustered based upon their liking and choices.

3. Land use: Used to identify areas with similar land used in GIS database, very useful to know the best use and the purpose the land serves such as near market, quality of land, suitable for construction or farming etc.
4. Search Engines: The search results often appear closest to the search history. It is done by making clusters with similar objects far from dissimilar objects.
5. Detection of Cancer Cells: Clustering algorithms are generally used in the detection of cancerous cells. It separates the dataset into clusters based upon the cancerous and non-cancerous cells.
6. Libraries: Used for categorizing various books based upon their genre.
7. City Planning: Used to group houses and identify their values based upon locality, geographical location etc.
8. Earthquake studies: By learning from the earthquake activity the dangerous zones can be determined.

3.6.3 Iterative Distance-Based Clustering

Iterative clustering is based upon the closeness of the centroid to other data points. K- Means is the commonly known algorithm that uses this clustering method. Here, the number of clusters is needed to be specified beforehand and therefore the prior knowledge of the dataset is necessary. All the instances are allocated to the cluster center that is nearest to them based upon the Euclidean distance metric. Coming down to the 'mean' part of the name of the k-Means algorithm, the mean or the centroid is selected for each and every cluster, which is the center value of that particular cluster [19]. The same process is iterated until the same data points are allocated to the clusters. When the cluster is stabilized it stays in that state as shown in Figure 3.13.

After the stabilization of the number of iterations, the total of the squared distance from all the points to the center of their clusters is minimized by assigning each point to the center of the nearest cluster. Always the global optimal clusters should be searched instead of local clusters i.e., the one with the least total squared distance. It is one of the most simple and effective method and has a linear complexity of the order of $O(n)$. It should be noted that the initial centers of the clusters must be chosen carefully. For selecting the best center, at first choose a random point having a uniform probability distribution throughout the entire space. The other seed is picked by checking the probability that is proportionate to the square of

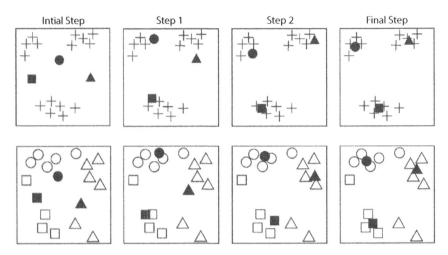

Figure 3.13 Iterative distance-based clustering [20].

the distance between the first and second point. Repeat the same procedure for all the clusters and it will improve the accuracy of the algorithm while increasing its speed as well. There is another method known as K-Median which is not as effective as K-Means, where the median vector is used for grouping. The advantage associated with this method is that it is less sensitive towards outliers. Its speed is quite low because sorting is necessary on each repetition while finding the median vector for big datasets [21].

3.7   Hierarchical Model

Hierarchical clustering is one of the most popular clustering methods. Its aim is to group objects with similar features into clusters in the form of a tree in such a way that the data points that were viewed to be most similar are held together in the branches [22]. There are basically two types of hierarchical clustering; Agglomerative clustering that is from leaves to root based upon bottom-up approach and Divisive clustering which is from root to leaves following top to bottom approach. Initially the entire observation is viewed as one single cluster and then segregation is done recursively as they go down the hierarchy [23]. The specification of the number of clusters expected is important as a termination constraint. Agglomerative clustering, also referred as Agglomerative Nesting (AGNES) is a clustering technique in which all the items are clusters in themselves making a total of n clusters. The, $n - 1$ clusters are left after joining two similar subjects.

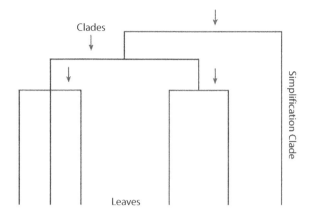

Figure 3.14 Parts of dendrogram [24].

Similarly, the procedure continues until all subjects are in the same cluster at the step $n - 1$ forming a hierarchical tree [24]. There are various linkage methods available for measuring distance among various clusters namely, Complete Linkage, Single-linkage, Average linkage, Centroid Linkage (calculating distance among centroids of two clusters). These are implemented in various standard software packages [25]. All this procedure is stored into dendrograms. It is a tree-like structure which reveals hierarchical relationships between the data points as shown in Figure 3.14. These are

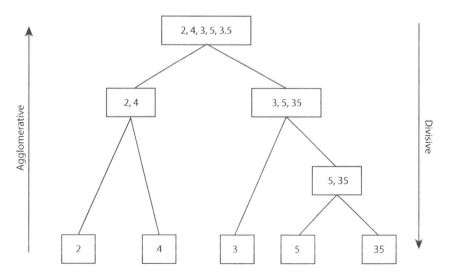

Figure 3.15 Types of hierarchical clustering [22].

generally in row or column format. The main parts of a dendrogram are Clades, which are nothing but branches arranged according to similarities (same height clades) and dissimilarities (varying height). Each clade is connected with one or many leaves.

Divisive clustering is the top to bottom approach and is also known as Divisive Analysis clustering (DIANA). In it, all the observations are assigned to a single cluster and then are divided into two non-similar clusters. This procedure is repeated recursively. The downside associated with this technique is that it is computation intensive. Both of the approaches can be visualized in Figure 3.15.

3.8   Density-Based Clustering

It is used to find clusters of arbitrary shape, by finding the dense region in the space separated by less dense region. Various clustering methods based on density can be understood by the basic three ways that are representative are OPTICS [26], DBSCAN [27]. These techniques are discussed in the subsequent sections.

3.8.1   DBSCAN

In it the algorithm finds a core data object having the maximum density around it. Then this data object and all the rest of the objects in the dense region are connected in order to form a cluster. For quantifying the neighbor of the data object, a parameter $\in$ is specified which is greater than zero. The neighbourhood of the object is the space between the radius whose center is the data object selected. Another parameter, MinPts, is used to check whether the surrounding is dense or not by specifying the threshold of the dense regions. An object is the core object when the minimum number of MinPts is contained by the $\in$-neighbourhood [19]. For finding the clusters in the dataset via DBSCAN, initially all objects are marked as unvisited. Then any point selected randomly is marked visited and checked whether $\in$-neighbourhood of that point consists of at least MinPts objects, if yes then it is a core object and a cluster is created for it and all the surrounding objects are added to it. Otherwise, it is marked as noise and other points are checked [19]. With proper setting of the parameters this method is quite adept at identifying the clusters of various shapes as shown in Figure 3.16. Its complexity is $O(n^2)$ and if positional reference is used then its computational complexity is $O(n \log \log n)$.

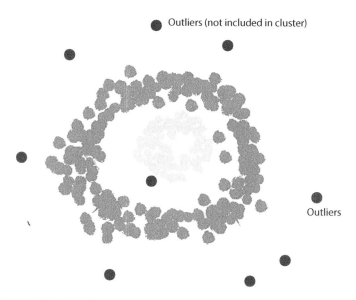

Figure 3.16 DBSCAN [19].

3.8.2 OPTICS

It overcomes the problems associated with the DBSCAN algorithm which is that their global parameters need to be user-defined. OPTICS does not produce data set clustering, it outputs cluster ordering, in which all the objects are linearly ordered [19]. Objects which are present in the clusters that are thicker are placed closer to each other. There is no need to specify a density threshold in OPTICS. All the information about the cluster, cluster visualization, cluster structure can be extracted by using the cluster ordering.

In order to obtain cluster ordering, clusters with the higher density are finished first, a few points are to be known about each object:

- The smallest value $\in'$ denotes the core distance of an object in such a way that least MinPts objects are attained by the $\in'$ neighborhood of the object. A minimum distance threshold $\in'$, makes the point a core object and in case it is not a core data point, the core distance is undefined.
- p density-reachable from q by minimum radius value, is the reachability distance to object from other point [19], where q, ought to be the core object and p is in its neighbourhood. The reachability distance from q to p is max of core distance

(q) and *dist.* (p, q) whereas, if q is not the core object, then the reachability-distance to p from q is undefined.

With regard to various core objects, an object p may have various reachability distances as it might be directly reachable from multiple core objects. The minimum reachability-distance of p is preferable as it will be at the closest distance from the dense cluster. OPTICS stores each object which has an appropriate reachability-distance. And computes ordering of all the objects. Objects which can be reached directly from the present core object are placed into a seed-list known as OrderSeeds which is for further augmentation [26]. As the current object, OPTICS starts with an arbitrary object from the input database, p and determines its core-distance, retrieves the ∈-neighbourhood of p, and marks the reachability-distance as *undefined*. The object, p, is then written as the output. If p is not the core object, then the next object in the defined list is selected whereas if it is the core object then in that case the reachability distance from p is updated. This procedure repeats until OrderSeeds is empty and input is fully consumed. The cluster ordering of the dataset can be visualized properly by representing it graphically. Its complexity is similar to DBSCAN i.e. $O(n \log \log n)$, when space index is used otherwise $O(n^2)$, where n denotes the number of objects.

3.9   Role of Machine Learning in Factories of the Future

The new era is all about the uprising of technology with the booming IT industry. The factories and manufacturing units have a lot of new data at their disposal which includes different formats, quality, and semantics such as sensor data from the production unit, parameters of the machine tools, environmental data and a lot more [28]. This availability of a big chunk of information is referred to as big data. This data helps in conducting a proper check over the improvement of the product and improving the procedure of manufacturing. This huge amount of data can even generate troublesome problems of various casualties. Machine learning and artificial intelligence will prove to be an appropriate tool for dealing with such situations with respect to manufacturing applications. The engagement of machine learning is spreading constantly over the past few decades. After many revolutions in various fields, came the fourth industrial revolution [29] by the use of machine intelligence that can be used for predicting future states and finding the best optimal solution for the systems, these intelligent systems are known as predictive manufacturing systems, PMS [30]. It promotes self-awareness, prediction, learning and maintenance of the machines. This provides transparency to the systems via which many probable risks and issues can be avoided within the machine [31]. Knowledge from the data can be extracted with the help of KDD, which is the discovering of knowledge in databases and procedure of finding and extracting knowledge from the data [32]. The application of machine learning is related to finding patterns in the dataset. It even improves the extracted knowledge as ML has the capability of finding new valuable information [28]. Machine learning also has utilization in optimization, controlling the quality and troubleshooting [33].

Machine learning techniques are believed to be the perfect tool for permanent quality improvement on large scale and complex processes [34]. With all the on-going development in the technology for smart manufacturing many new issues are arising for the future development. Despite all the enhancement in research and technology there is a huge need of practically addressing the implementation of the plans and strategies. Smart manufacturing is being carried out in conceptual approach designing sets whereas future approaches need to look forward towards application strategies for smart manufacturing including plans, designs, maintenance, manufacturing and development. Application guidelines and reference models are basic necessities for the application of the technologies. It will help in saving cost and increasing the production while creating new values that generates and serves the purpose of the society. Human and society thinking should go hand in hand for its success. Machine learning will not only construct simple intelligent machines but even develop a continuous growth engine for manufacturing leading to sustainable development [29].

3.10   Identification of the Probable Customers

This section contains a practical implementation of one of the machine learning algorithm that aims at solving the classification problem of identifying potential customers who can buy the product based on various features using machine learning algorithm. The algorithm that has been used to detect the potential customers in this work is K-Nearest Neighbors (KNN). In this section not only a solution to the problem has been provided but also a complete description of the model pipeline has been outlined. To begin with, KNN is an unsupervised machine learning algorithm that classifies a particular data entry on the basis of the number of data points of various classes surrounding it. The data point under observation is assigned a class that is based on the distance among the current data point and all the other data points that are around it. Various distances metrics that can be used for finding the association are Euclidean distance, Minkowski distance, or Manhattan distance. All those data points that have minimum distance among themselves form a cluster. Similarly other clusters are created where each cluster is different from other clusters. In KNN, *k* represents the number of neighbors that are used to find the class of the data point. Following is the code for KNN implementation to detect the customers.

```
import pandas as pd
import numpy as np
import matplotlib.pyplot as plt
from sklearn import metrics
from sklearn.preprocessing import StandardScaler,
OneHotEncoder
from sklearn.model_selection import train_test_
split, StratifiedKFold, KFold

from sklearn.compose import ColumnTransformer
from sklearn.metrics import accuracy_score
from sklearn.neighbors import KNeighborsClassifier
from matplotlib.colors import ListedColormap
import time
t1 = time.time()

#load data

df = pd.read_csv('Social_Network_Ads.csv')
X = df.iloc[:,0:2].values
y = df.iloc[:,2].values

###Split data
X_dataset, X_test, y_dataset, y_test = train_test_
split(X,y,test_size=0.2,random_state=0)
sc = StandardScaler()
L = []

classifier = KNeighborsClassifier(n_neighbors = 5,
metric = 'minkowski', p=2)
kf = KFold(n_splits=5, shuffle=True, random_state=0)
```

```
for train_index, val_index in kf.split(X_dataset,
y_dataset):
    X_train, y_train = X_dataset[train_index,:],
y_dataset[train_index]
    X_val,   y_val   =   X_dataset[val_index,:],
y_dataset[val_index]
    X_train = sc.fit_transform(X_train)
    X_val = sc.transform(X_val)
    classifier.fit(X_train, y_train)
    y_pred = classifier.predict(X_val)
    acc = accuracy_score(y_val, y_pred)
    print('Accuracy = {}'.format(acc))
    L.append(acc)
print('Mean accuracy = {}'.format(np.mean(L)))
#Test set prediction
X_test = sc.transform(X_test)
y_pred = classifier.predict(X_test)
acc = accuracy_score(y_test, y_pred)
print('Accuracy on test set = {}'.format(acc))

#Visualization

X1 = np.arange(np.min(X_train[:,0])-1, np.max(X_
train[:,0])+1, step = 0.01)
X2 = np.arange(np.min(X_train[:,1])-1, np.max(X_
train[:,1])+1, step = 0.01)
xv, yv = np.meshgrid(X1, X2, indexing='ij')
plt.contourf(xv,    yv,    classifier.predict(np.
asarray([xv.ravel(),         yv.ravel()]).T).
reshape(xv.shape),   alpha   =0.4,   cmap   =
ListedColormap(('red','green')))
for index, values in enumerate(np.unique(y_test)):
    plt.scatter(X_test[y_test==index,0],
X_test[y_test==index,1],         c=ListedColor-
map(('red','green'))(values),label = values)
plt.title('K-NN Test Results')
plt.xlabel('Age')
plt.ylabel('Salary')
plt.legend()
t2 = time.time()
print('Time (in secs): {}'.format(t2-t1))
plt.show()
```

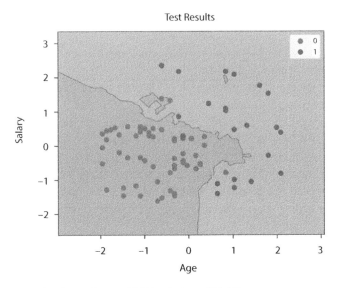

Figure 3.17 Results obtained using KNN technique (KFold).

In this work, the value for k is chosen to be 5 as it gives the maximum performance in terms of accuracy. As in this work, Kfold cross validation is used where the number of folds have also been chosen to be 5, so we obtain 5 different accuracies of 0.906, 0.94, 0.89, 0.86, 0.89 but overall KNN algorithm achieves an average accuracy of 89.26% on the validation set and an accuracy of 94.89% on the test set. Visual quality results are shown in Figure 3.17.

Visual representation shows that KNN performs very effective classification of the buying and non-buying customers.

3.11   Conclusion

Machine learning is a fast evolving field and is applied in nearly every field of study nowadays. It enables solving really complex problems that are too difficult for humans and are very time consuming. Machine learning techniques work by analyzing different kinds of patterns present in the data and making predictions based on the patterns. In this chapter, various types of machine learning are discussed in detail along with the distance metrics for the evaluation of the models. Besides supervised learning, a detailed discussion has been provided on unsupervised techniques such as clustering techniques. Dimensionality plays a very crucial role in the performance of the models and for the same a detailed analysis has been done about reducing the dimensionality of the data in order to increase the precision of the models. Various applications and the role of machine learning have also been provided.

References

1. Sodhi, P., Awasthi, N., Sharma, V., Applications, C., Introduction to machine learning and its basic application in Python, *Proceedings of 10th International Conference on Digital Strategies for Organizational Success*, pp. 1354–1375, 2017.
2. Khanum, M., A survey on unsupervised machine learning algorithms for automation, classification and maintenance. *Int. J. Comput. Appl.*, 119, 13, 34–39, June, 2015.
3. Berry, M.W., *Supervised and unsupervised learning for data science*, Springer Publishing Company, Incorporated, New York, USA, January, 2020.
4. Liu, Y. *et al.*, Modification on the tribological properties of ceramics lubricated by water using fullerenol as a lubricating additive. *Sci. China Technol. Sci.*, 55, 9, 2656–2661, Sep. 2012.
5. Van Otterlo, M. and Wiering, M., Chapter 1: Reinforcement learning and Markov decision, in: *Reinf. Learn. State-of-the-Art*, pp. 3–42, 2012.
6. Barto, A.G., Sutton, R.S., Anderson, C.W., Looking back on the actor–critic architecture, in: *IEEE Transactions on Systems, Man, and Cybernetics: Systems*, vol. 51, 1, pp. 40–50, Jan. 2021.
7. Lee, D., Seo, H., Jung, M.W., Neural basis of reinforcement learning and decision making. *Annu. Rev. Neurosci.*, 35, 1, 287–308, Jul. 2012.
8. *Introduction to dimensionality reduction*, GeeksforGeeks, Noida, India, 2021, https://www.geeksforgeeks.org/dimensionality-reduction/ (accessed Jun. 28, 2021).
9. Ramadevi, G. and Usharani, K., Study on dimensionality reduction techniques and applications. *Publ. Probl. Appl. Eng. Res.*, 04, 1, 134–140, 2013.
10. *Using Linear Discriminant Analysis (LDA) for data explore: Step by step*, Blog, Palma, Spain, 2021, https://www.apsl.net/blog/2017/07/18/using-linear-discriminant-analysis-lda-data-explore-step-step/ (accessed Jun. 28, 2021).
11. Weller-fahy, D.J., Borghetti, B.J., Sodemann, A.A., Within network intrusion anomaly detection. *IEEE Commun. Surv. Tutorials*, 17, 1, 70–91, 2015.
12. Grootendorst, M., *9 Distance measures in data science*, Towards Data Science, Ontario, Canada, pp. 1–13, 2021, Accessed: Jun. 07, 2021. [Online]. Available: https://towardsdatascience.com/9-distance-measures-in-data-science-918109d069fa.
13. Banerjee, W. and Vidhya, A., *Role of distance metrics in machine learning*, Towards Data Science, Ontario, Canada, Jun. 12, 2020, https://medium.

com/analytics-vidhya/role-of-distance-metrics-in-machine-learning-e43391a6bf2e (accessed Jun. 08, 2021).

14. Pandit, S. and Gupta, S., A comparative study on distance measuring approaches for clustering. *Int. J. Res. Comput. Sci.*, 2, 1, 29–31, Dec. 2011.

15. Abu Alfeilat, H.A., Hassanat, A.B.A., Lasassmeh, O., Tarawneh, A.S., Alhasanatm M.B., Wyal Salman, H.S., Prasath, V.B.S., Effects of distance measure choice on KNN classifier performance. A review. *Big Data*, 7, 4, 221–248.

16. Bora, D.J., A comparative study between fuzzy clustering algorithm and hard clustering a comparative study between fuzzy clustering algorithm and hard clustering algorithm. *International Journal of Computer Trends and Technology (IJCTT)*, 10, 2, 108–113, April 2014.

17. Peters, G., Crespo, F., Lingras, P., Weber, R., Soft clustering – fuzzy and rough approaches and their extensions and derivatives. *Int. J. Approx. Reason.*, 54, 2, 307–322, 2013.

18. Ester, M., Kriegel, H.-P., Sander, J., Xu, X., A density-based algorithm for discovering clusters in large spatial databases with noise, in: *Proceedings of the 2nd International Conference on Knowledge Discovery and Data Mining*, pp. 226–231, 1996, Accessed: May 14, 2021. [Online]. Available: www.aaai.org.

19. Han, J., Kamber, M., Pei, J., Cluster analysis, (Third Edition), The Morgan Kaufmann Series in data management systems, in: *Data Mining*, pp. 443–495, Elsevier, Burlington, Massachusetts, United States, 2012.

20. Witten, I.H., Frank, E., Hall, M.A., Pal, C.J., *Data Mining* (Fourth Edition), The Morgan Kaufmann Series in Data Management Systems, Elsevier, Burlington, Massachusetts, United States, 2017.

21. Seif, G., *The 5 clustering algorithms data scientists need to know*, Towards Data Science, Ontario, Canada, Feb. 05, 2018, https://towardsdatascience.com/the-5-clustering-algorithms-data-scientists-need-to-know-a36d136ef68 (accessed Jun. 08, 2021).

22. Tullis, T. and Albert, B., *Measuring the user experience*, Burlington, Massachusetts, United States, Elsevier, 2013.

23. Zhang, Z., Murtagh, F., Van Poucke, S., Lin, S., Lan, P., Hierarchical cluster analysis in clinical research with heterogeneous study population: Highlighting its visualization with R. *Ann. Transl. Med.*, 5, 4, 75–75, Feb. 2017.

24. Bergman, L.R. and Magnusson, D., Person-centered research, in: *International Encyclopedia of the Social & Behavioral Sciences*, pp. 11333–11339, Oxford, UK, Elsevier, 2001.

25. Bunge, J.A. and Judson, D.H., Data mining, in: *Encyclopedia of Social Measurement*, pp. 617–624, Elsevier, Newyork, United States, 2005.

26. Ankerst, M., Breunig, M.M., Kriegel, H.-P., Sander, J., OPTICS, in: *Proceedings of the 1999 ACM SIGMOD International Conference on Management of Data - SIGMOD '99*, pp. 49–60, 1999.

27. Hyde, R. and Angelov, P., Data density based clustering. *2014 14th UK Work. Comput. Intell. UKCI 2014 – Proc.*, Ddc, 2014.

28. Wuest, T., Weimer, D., Irgens, C., Thoben, K.-D., Machine learning in manufacturing: Advantages, challenges, and applications. *Prod. Manuf. Res.*, 4, 1, 23–45, Jan. 2016.

29. Liao, Y., Deschamps, F., de F. R. Loures, E., Ramos, L.F.P., Past, present and future of Industry 4.0 - a systematic literature review and research agenda proposal. *Int. J. Prod. Res.*, 55, 12, 3609–3629, Jun. 2017.

30. Nikolic, B., Ignjatic, J., Suzic, N., Stevanov, B., Rikalovic, A., Predictive manufacturing systems in Industry 4.0: Trends, benefits and challenges, in: *Annals of DAAAM & Proceedings*, pp. 796-802, 2018

31. Lee, J., Lapira, E., Bagheri, B., Kao, H., Recent advances and trends in predictive manufacturing systems in big data environment. *Manuf. Lett.*, 1, 1, 38–41, Oct. 2013.

32. Sugawara, E. and Nikaido, H., Properties of AdeABC and AdeIJK efflux systems of *Acinetobacter baumannii* compared with those of the AcrAB-TolC system of Escherichia coli. *Antimicrob. Agents Chemother.*, 58, 12, 7250–7257, Dec. 2014.

33. Alpaydin, E., *Introduction to machine learning*, MIT Press, Massachusetts, United States, 2004.

34. Pham, D.T. and Afify, A.A., Machine-learning techniques and their applications in manufacturing, in: *Proceedings of the Institution of Mechanical Engineers, Part B: Journal of Engineering Manufacture*, 219, 5, 395–412 2005.

4

Understanding Neural Networks

Er. Lal Chand*, Sikander Singh Cheema and Manpreet Kaur

*Department of Computer Science and Engineering, Punjabi University,
Patiala, Punjab, India*

Abstract

Neural Networks have seen an explosion of interest over the last few years. The primary appeal of neural networks is their ability to emulate the brain's pattern recognition skills. The sweeping success of neural networks can be attributed to some key factors. This paper explains the architecture of neural networks and also enlightens how neural networks are being successfully applied an extraordinary range of problem domain [1].

Neural Networks (NN) are important data mining tools used for classification and clustering. It is an attempt to build machines that will mimic brain activities and be able to learn. Neural Networks usually learn by examples. If NN is supplied with enough examples, it should be able to perform classification and even discover new trends or patterns in data. Basic NN is composed of three layers, input, output, and hidden layers. Each layer can have number of nodes and nodes from input layers are connected to the nodes from hidden layer. Nodes from hidden layers are connected to the nodes from output layer. Those connections represent weights between nodes [2].

Keywords: Activation function, machine learning, convolution neural network (CNN), RNN, back propagation

4.1 Introduction

A neural network is a group of algorithms that certify the underlying relationship in a set of data similar to the human brain. The neural network helps to change the input so that the network gives the best result without

---

*Corresponding author: lc.panwar5876@gmail.com

Chandan Deep Singh and Harleen Kaur (eds.) Factories of the Future: Technological Advancements in the Manufacturing Industry, (83–102) © 2023 Scrivener Publishing LLC

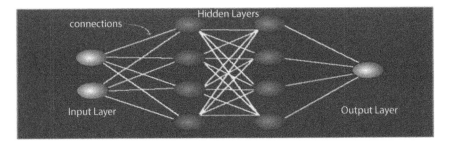

Figure 4.1 Neural network [4].

redesigning the output procedure. An example of neural network is shown in Figure 4.1.

The human brain is the most complex organ in the human body. It helps us think, understand, and make decisions. The secret behind its power is a neuron. Ever since the 1950s, scientists have been trying to mimics the functioning of a neuron and use it to make smarter and better robots. After a lot of trial and error, humans finally created a computer that could recognize human speech. It was only after the year 2000 that people were able to master deep learning (a subset of AI) that was able to see and distinguish between various images and videos [3, 4].

A neural network works similarly to the human brain's neural network. A "neuron" in a neural network is a mathematical function that collects and classifies information according to a specific architecture. The network bears a strong resemblance to statistical methods such as curve fitting and regression analysis [6].

- They are used in a variety of applications in financial services, from forecasting and marketing research to fraud detection and risk assessment.
- Use of neural networks for stock market price prediction varies.

4.2 Components of Neural Networks

There are different types of neural networks but they always consist of the same components: neurons, synapses, weights, biases, and functions.

4.2.1   Neurons

A neuron or a node is a basic unit of neural networks that receives information, performs simple calculations, and passes it further. All neurons in a network are divided into three groups:

- Input neurons that receive information from the outside world.
- Hidden neurons that process that information;
- Output neurons that produce a conclusion [5].

The typical nerve cell of the human brain comprises of four parts –

- **Function of Dendrite**
  It receives signals from other neurons.
- **Soma (cell body)**
  It sums all the incoming signals to generate input.
- **Axon Structure**
  When the sum reaches a threshold value, the neuron fires, and the signal travels down the axon to the other neurons.
- **Synapses Working**
  The point of interconnection of one neuron with other neurons [21]. The amount of signal transmitted depends upon the strength (synaptic weights) of the connections. Different parts and their functions of neurons are shown below in Figure 4.2.

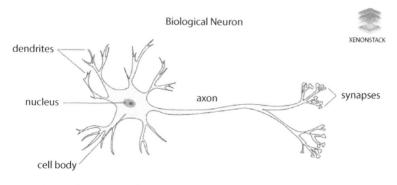

Figure 4.2 Parts of neurons and their functions [5].

4.2.2   Synapses and Weights

A synapse is what connects the neurons like an electricity cable. Every synapse has a weight. The weights also add to the changes in the input information. The results of the neuron with the greater weight will be dominant in the next neuron, while information from less 'weighty' neurons will not be passed over. We can say that the matrix of weights governs the whole neural system [27]. How information is passed to the next neuron on the basis of weights is shown in the Figure 4.3.

4.2.3   Bias

A bias neuron allows for more variations of weights to be stored. Biases add richer representation of the input space to the model's weights role by making it possible to move the activation function to the left or right on the graph. An example of biased neuron is shown below in the Figure 4.4.

4.2.4   Architecture of Neural Networks

A single neuron may takes more then one inputs perform some operations to generate an output. The architecture of multilayer neural networks is illustrated in the Figure 4.5.

- **Input layer** – It contains those units (Artificial Neurons) which receive input from the outside world on which the network will learn, recognize about, or otherwise process.

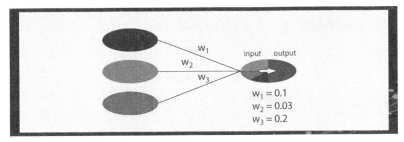

Figure 4.3 Synapses and weights [5].

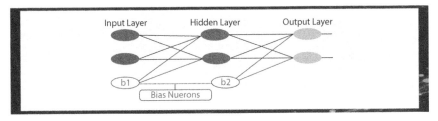

Figure 4.4 Bias [5].

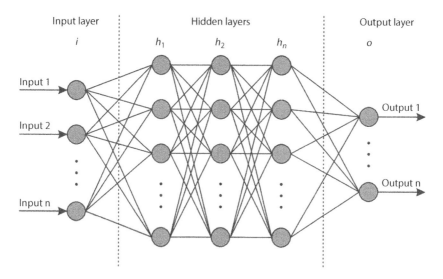

Figure 4.5 Architecture of neural networks [17].

> **Output layer** – It contains units that respond to the information about how it learns any task.
> - **Hidden layer** – These units are in between input and output layers. The hidden layer's job is to transform the input into something that the output unit can use somehow.

4.2.5   How Do Neural Networks Work?

- Information is fed into the input layer which transfers it to the hidden layer.
- The interconnections between the two layers assign weights to each input randomly.
- A bias added to every input after weights is multiplied with them individually.
- The weighted sum is transferred to the activation function.
- The activation function determines which nodes it should fire for feature extraction.
- The model applies an application function to the output layer to deliver the output.
- Weights are adjusted, and the output is back-propagated to minimize error.

Working of neural networks can be understood from Figure 4.6 shown below.

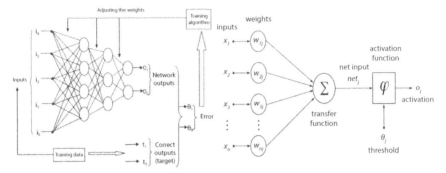

Figure 4.6 Working of neural networks [22].

4.2.6 Types of Neural Networks

- Convolution Neural Network (CNN)
- Recurrent Neural Network (RNN)
- Artificial Neural Network (ANN) [11]

4.2.6.1 Artificial Neural Network (ANN)

- An artificial neural network represents the structure of a human brain modeled on the computer. It consists of neurons and synapses organized into layers. As shown in the Figure 4.7.
- ANN can have millions of neurons connected into one system, which makes it extremely successful at analyzing and even memorizing various information [12].

Here is a video for those who want to dive deeper into the technical details of how artificial neural networks work.

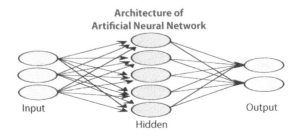

Figure 4.7 Artificial neural network [18].

4.2.6.2   Recurrent Neural Network (RNN)

RNN is a type Neural Network where the output from previous step is fed as input to the current step. The artificial neural network is used in text-to-speech conversion technology. In traditional neural networks, all the inputs and outputs are independent of each other, but in cases like when it is required to predict the next word of a sentence, the previous words are required and hence there is a need to remember the previous words. Thus RNN came into existence, which solved this issue with the help of a Hidden Layer.

Recurrent neural networks are widely used in natural language processing and speech recognition [20].

4.2.6.3   Convolutional Neural Network (CNN)

This type of neural network uses a variation of the multilayer perceptions. Convolution neural networks contain single or more than one layer that can be pooled or entirely interconnected. They show good results in paraphrase detection and semantic parsing. They are applied in image classification and signal processing.

This network consists of one or multiple convolutional layers. The convolutional layer present in this network applies a convolutional function on the input before transferring it to the next layer. Due to this, the network has fewer parameters, but it becomes more profound. CNNs are widely used in natural language processing and image recognition. Convolutional Neural Networks (CNN) is a variant of Multi-Layer Perception (MLPs)

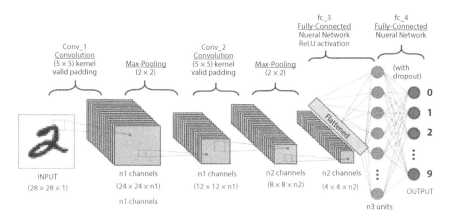

Figure 4.8 Convolutional neural network [7].

which are inspired from biology. Convolutional neural networks are designed to process two-dimensional (2-D) image as shown in the Figure 4.8. The network consists of three types of layers namely convolution layer, sub sampling layer and the output layer.

4.2.7 Learning Techniques in Neural Network

- **Supervised Learning**
  In this learning, the training data is input to the network, and the desired output is known weights are adjusted until production yields desired value.
- **Unsupervised Learning**
  It can be Use the input data to train the network whose output is known. The network classifies the input data and adjusts the weight by feature extraction in input data.
- **Reinforcement Learning**
  In this learning technique, the output value is unknown, but the network provides feedback on whether the output is right or wrong. It is also known as Semi-Supervised Learning.
- **Offline Learning**
  The weight vector adjustment and threshold adjustment are made only after the training set is shown to the network. It is also known as Batch Learning.
- **Online Learning**
  In this learning technique, the adjustment of the weight and threshold is made after presenting each training sample to the network.

4.2.8 Applications of Neural Network

Applications of neural network can be described as follows:

- **Handwriting Recognition**
  Neural networks are used to convert handwritten characters into digital characters that a machine can recognize.
- **Stock-Exchange Prediction**
  The stock exchange is difficult to track and difficult to understand. Many factors affect the stock market. A neural network can examine a lot of factors and predict the prices daily, which would help stockbrokers.

- **Traveling Issues of Sales Professionals**
  This type refers to finding an optimal path to travel between cities in a particular area. Neural networks help solve the problem of providing higher revenue at minimal costs.
- **Image Compression**
  The idea behind the data compression neural network is to store, encrypt, and recreate the actual image again. We can optimize the size of our data using image compression neural networks. It is the ideal application to save memory and optimize it [4, 5].

4.2.9   Advantages of Neural Networks

- **Store information on the entire network**
  It happens in traditional programming where information is stored on the network and not on a database. If a few pieces of information disappear from one place, it does not stop the whole network from functioning.
- **The ability to work with insufficient knowledge**
  After the training of ANN, the output produced by the data can be incomplete or insufficient. The importance of that missing information determines the lack of performance.
- **Good fault tolerance**
  The output generation is not affected by the corruption of one or more than one cell of artificial neural network. This makes the networks better at tolerating faults.
- **Gradual corruption**
  Indeed a network experiences relative degradation and slows over time. But it does not immediately corrode the network.
- **Ability to train machine**
  ANN learns from events and makes decisions through commenting on similar events.
- **The ability of parallel processing**
  These networks have numerical strength which makes them capable of performing more than one function at a time [13].

4.2.10   Disadvantages of Neural Network

1. The most important disadvantages of neural networks are their black-box nature.

2. Neural networks usually require much more data than traditional algorithms, as in at least thousands if not millions of labeled samples.
3. Neural networks are very complex in computing terms other than traditional algorithms.
4. The duration of the neural network is unknown.
5. Hardware dependence.
6. Unexplained behavior of the network.
7. Determination of proper network structure.
8. The difficulty of showing the problem of the network.

4.2.11   Limitations of Neural Networks

- Neural Network learning algorithm are inductive, requiring large amount of data, whereas strategic decision making deals with unique and non-routine types of decision making.
- Neural network do not provide explanations for their decisions.
- Neural Network decisions are not supported by significant tests, hence low validity.

4.3   Back-Propagation

Back-propagation is a short form for "backward propagation of errors." It is a standard method of training artificial neural networks. This method helps to calculate the gradient of a loss function with respects to all the weights in the network [8].

4.3.1   Working of Back-Propagation

1. Inputs X, arrive through the pre-connected path
2. Input is modeled using real weights W. The weights are usually randomly selected.
3. Calculate the output for every neuron from the input layer, to the hidden layers, to the output layer.
4. Calculate the error in the outputs.

$$\text{Error}_B = \text{Actual Output} - \text{Desired Output}$$

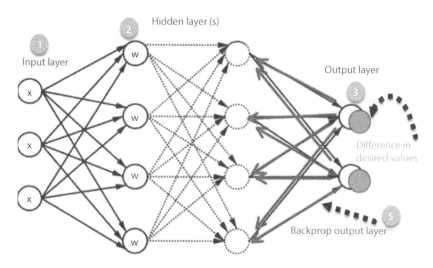

Figure 4.9 Backpropagation [8].

5. Travel back from the output layer to the hidden layer to adjust the weights such that the error is decreased.

Keep repeating the process until the desired output is achieved as shown above in the Figure 4.9 [8].

4.3.2   Types of Back-Propagation

Back-propagation network are of two types.

4.3.2.1   Static Back-Propagation

It is a type of network which produces a mapping of static input for static output. It is used to solve static classification problems.
Like: optical character recognition.

4.3.2.2   Recurrent Back-Propagation

It is fed forward until a fixed value is achieved. After that the error is computed and propagated backward.
The main difference between both of these methods is:
The mapping is rapid in static back-propagation while it is non-static in recurrent back-propagation [8].

4.3.2.3   Advantages of Back-Propagation

Main advantages of Back propagation are:

- It is fast, simple and easy to program
- It has no parameters to tune apart from the numbers of input
- It is a flexible method as it does not require prior knowledge about the network
- It is a standard method that generally works well
- It does not need any special mention of the features of the function to be learned.

4.3.2.4   Disadvantages of Back-Propagation

- The actual performance of this technique on a specific problem is dependent on the input data.
- It can be quite sensitive to noisy data.
- We need to use the matrix based approach for back-propagation instead of mini-batch [8].

4.4   Activation Function (AF)

Definition of activation function:-It decides, whether a neuron should be activated or not by calculating weighted sum and further adding bias with it. The aim of the activation function is to introduce non-linearity into the output of a neuron [10, 15].

4.4.1   Sigmoid Active Function

1. It is a function which as 'S' shaped graph.
2. **Equation:**

$$A = 1/(1 + e^{-Y})$$

3. **Value Range:** 0 to 1
4. **Uses:** It is used in output layer of a binary classification, where result is either 0 or 1, as value for sigmoid function lies between 0 and 1 only, result can be predicted easily to be *1* if value is greater than **0.5** and *0* otherwise.

Sigmoid function is a widely used activation function.
It is defined as graphically as shown in the Figure 4.10.

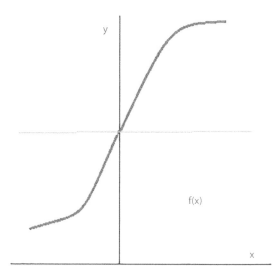

Figure 4.10 Graphical representation of sigmoid active function [15].

4.4.1.1   Advantages

- It can be Easy to understand and apply.
- It can be Easy to train on dataset.
- This is a smooth function and is continuously differentiable.
- It is non-linear. The output is non-linear as well.
- It can be easy to compute differential.

4.4.1.2   Disadvantages

- Vanishing gradient problem, the output is not zero centered.
- It can be slow convergence.

4.4.2   RELU Activation Function

It stands for Rectified linear unit. It can be implemented in hidden layer of neural network and the graphical representation of RELU active function is shown below in the Figure 4.11.

- **Equation:-A(X) = *max (0,x).*** It gives an output x if x is positive and 0 otherwise.
- **Value Range :-** [0, inf)

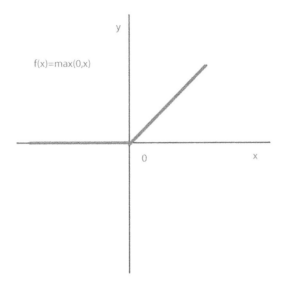

Figure 4.11 Graphical representation of RELU active function [16].

- **Nature :-** Non-Linear
- **Uses:-**RELU is less computationally expensive than TANH and sigmoid because it involves simpler mathematical operations. At a time only a few neurons are activated making the network sparse making it efficient and easy for computation.

It is defined as graphically.

4.4.2.1 Advantages

- Avoids and rectifies vanishing gradient problems.
- RELU learns much faster than sigmoid and TANH function.

4.4.2.2 Disadvantages

- It can only be used with a hidden layer.
- It can be hard to train on small datasets need much data for learning non-linear behavior.
- Non-differentiable at zero and RELU is unbounded.

4.4.3 TANH Active Function

It is also known as Tangent Hyperbolic Function. It is mathematically shifted version of the sigmoid function and the graphical representation of TANH function is shown below in the Figure 4.12.

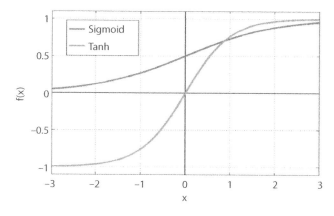

Figure 4.12 Graphical representation of TANH function [16].

- **Equation:** $f(x) = \tanh(x) = 2(1+e-2x)-1$
- **Range:** -1 to +1
- **Nature:** non-linear
- **Uses:** -It can be used in hidden layers of Neural Network. The value of this function is -1 to 1 that means the hidden layer comes out be 0 or very close to it. It helps in centering the data by bringing mean close to 0.

4.4.3.1   Advantages

- The output is zero centered.
- It is continuous and differentiable at all points.
- The function as you can see is non-linear, so we can easily back-propagate the errors.

4.4.3.2   Disadvantages

- Vanishing gradient problem, hard to train on small datasets.
- The gradients are low [15].

4.4.4   Linear Function

- **Equation:** y = ax
- **Range:** -inf to + inf
- **Uses:** It is used in always one place i.e. output layer.

4.4.5   Advantages

- The linear function might be ideal for simple tasks where interpretability is high desired.

4.4.6   Disadvantages

- The derivative of a linear function (i.e. 'a') is constant. It does not depend upon the input value x. This means that every time we do a back propagation, the gradient would be same.
- If each layer has a linear transformation, no matter how many layers we have the final output is nothing but a linear transformation of the input.

4.4.7   Softmax Function

It is type of sigmoid function but it is handy when we are trying to handle classification problems.

- **Nature:** Non-linear
- **Uses:** It is used when trying to handle multiple classes. The outputs for each class between 0 and 1and would also divide by the sum of the outputs.

4.4.8   Advantages

It is useful for output neurons. It is used only for the output layer, for neural networks that need to classify inputs into multiple categories.

4.5   Comparison of Activation Functions

The above discussed activation functions Sigmoid, RELU, TANH, Linear and Softmax are compared on the basis of four attributes Range, Nature, Mathematical Equation and Uses is shown in the form of Table 4.1.

Table 4.1 Comparison of activation functions [15].

Activation functions	Uses	Range	Nature	Equation
Sigmoid Function	It is used in output layer of a binary classification.	0 to 1	Non-Linear	$Y=1/(1+e^\wedge(-x))$

(Continued)

Table 4.1 Comparison of activation functions [15]. (*Continued*)

Activation functions	Uses	Range	Nature	Equation
RELU Function	it involves simpler mathematical operations.	0 to infinite	Linear	{xi if x >= 0 0 if x <=0}
TANH Function	Usually used in hidden layers of a neural network.	-1 to 1	Non-Linear	y = tanh(x)
Linear Function	It is used at just one place i.e. output layer.	Inf to + inf	Linear	Linear function has the equation similar has to straight line i.e. **y = ax.**
Softmax Function	It is used when trying to handle multiple classes.	0 and 1	Non-Linear	$$S(y)_i = \frac{\exp(y_i)}{\sum\limits_{j=1}^{n} \exp(y_j)}$$

4.6 Machine Learning

It is a method of data analysis that can be generate analytical model building. It is branch of artificial intelligence based on the idea that system can learn from data, identify patterns and make decisions with minimal human interventions.

It is the field of study that gives computers to capability to learn without being explicitly programmed. It is one of the most exciting technologies that one would have ever come across. It gives the computer that makes it

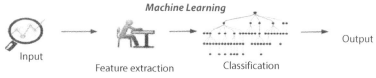

Traditional machine learning uses hand-crafted features, which is tedious and costly to develop.

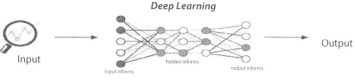

Deep learning learns hierarchical representation from the data itself, and scales with more data.

Figure 4.13 Machine learning [14].

more similar to human The basic concept of machine learning and deep learning are shown above in the Figure 4.13.

It is actively being used today, perhaps in many more places that one would expect [9].

4.6.1   Applications of Machine Learning

- **Image Recognition**
  It is one of the most common applications of machine learning. Machine Learning is used to identify objects, persons, places, digital images etc. It is based on the Facebook project named "deep face", which is responsible for face recognition and person identification in the picture [19].
- **Self Driving Cars**
  It plays a significant role in self-driving cars. Tesla, the most popular car manufacturing company is working on self-driving car. It is using unsupervised learning method to train the car models to detect people and objects while driving.
- **Stock Marketing Trading**
  It is widely used in stock market trading. In the stock market, there is always a risk of up and downs in shares, so for this machine learning's long short term memory neural network is used for the prediction of stock market trends.
- **Medical Diagnosis**
  It is used for diseases diagnoses. Medical technology is growing very fast and able to build 3d models that can predict the exact position of lesions of brain.
  It helps in finding brain tumors and other brain related diseases easily.
- **Automatic Language Translation**
  The technology is behind the automatic translation is a sequence learning algorithm , which is used with image recognition and translate the text from one language to another language [24].

4.7   Conclusion

The computing world has a lot to gain from neural networks. Their ability to learn by example makes them very flexible and powerful. Furthermore there is no need to devise an algorithm in order to perform a specific task, i.e. there is no need to understand the terminal mechanisms of that task. Perhaps the most exciting aspect of neural networks is the possibility that some day 'conscious' networks might be produced. There are a number of scientists arguing that consciousness is a 'mechanical' property and that 'conscious' neural networks are a realistic possibility.

References

1. https://www.researchgate.net/publication/287410862_Artificial_neural_networks_for_e-NOSE_A_review.
2. https://www.ijert.org/neural-networks-in-data-mining.
3. Sabour, R. and Amiri, A., Comparative study of ANN and RSM for simultaneous optimization of multiple targets in fenton treatment of landfill leachate. *Waste Manage.*, 65, 54–62, July, 2017. [26].
4. https://www.simplilearn.com/tutorials/deep-learning-tutorial/what-is-neural-network.
5. https://www.google.com/search?q=introduction+to+neural+networks&sxsrf=ALiCzsZqAzKoSdZP3Fm4qEqUeNPzm2sgUQ:1653556505681&source=lnms&tbm=isch&sa=X&ved=2ahUKEwjj_4uB6vz3AhVq7HMBHTR0BQIQ_AUoAnoECAEQBA&biw=1366&bih=583&dpr=1#imgrc=CjorXvyAYo–aM.
6. http://pages.cs.wisc.edu/~bolo/shipyard/neural/local.html.
7. A comprehensive guide to Convolutional Neural Networks – the ELI5 way.
8. Johnson, D., Back propagation Neural Network: What is backpropagation algorithm in machine learning?, Updated October 6, 2021. https://www.guru99.com/backpropogation-neural-network.html
9. https://www.suse.com/suse-defines/definition/machine-learning/#:~:text=Machine%20learning%20also%20refers%20to,decisions%20with%20minimal%20human%20intervention.
10. https://media.geeksforgeeks.org/wp-content/uploads/1-106.png11.
11. https://www.analyticsvidhya.com/blog/2020/02/cnn-vs-rnn-vs-mlp-analyzing-3-types-of-neural-networks-in-deep-learning/.
12. https://www.folio3.ai/blog/advantages-of-neural-networks/.
13. https://www.google.com/search?q=machine+learning&bih=583&biw=1366&hl=enGB&sxsrf=ALiCzsZCST2SFPe3eZ9YeTRrPGdBCX-EvuQ:1653556746759&source=lnms&tbm=isch&sa=X&ved=2ahUKEwjyrIb06vz3AhWeUWwGHT8cBmsQ_AUoAnoECAIQBA#imgrc=kpy_4f8-tyVkZM.
14. https://www.google.com/search?q=Fig.4.3.1:+Graphical+Representation+of+Tanh+Function&sxsrf=ALiCzsarVb3kZuByhvwtci-U884MapkezBQ:1653556886970&source=lnms&tbm=isch&sa=X&ved=2ahUKEwi28_O26_z3AhV0TGwGHWKUBEEQ_AUoAXoECAEQAw#imgrc=C2opUvDMbgO99M.

15. https://www.google.com/search?q=convolutional+neural+netw
    ork&sxsrf=ALiCzsaeNPJXAdRwbVY22IZZ8ROH9FLxMA:16535
    57584535&source=lnms&tbm=isch&sa=X&ved=2ahUKEwjGjMS
    D7vz3AhWcUGwGHUmsCcEQ_AUoAXoECAEQAw&biw=13
    66&bih=583&dpr=1#imgrc=NF1b0qO0_JYxrM.
16. https://www.google.com/search?q=artificial+neural+network&sxsrf=ALiCz
    sbufhQD1xkiMcX4pvfP_YDBXEZh_A:1653557781075&source=lnms&tbm=
    isch&sa=X&ved=2ahUKEwjB_J_h7vz3AhUuUGwGHbASAC8Q_
    AUoAXoECAMQAw#imgrc=qOJTrNxztXCTTM.
17. https://www.guru99.com/backpropogation-neural-network.html.
18. https://www.google.com/search?q=architecture+of+neural+network&sx-
    srf=ALiCzsZq7VEG4DR1nASd7KvNIanHOIGUyw:1653558184947&-
    source=lnms&tbm=isch&sa=X&ved=2ahUKEwiYtOqh8Pz3AhX-
    XSmwGHdfRCscQ_AUoAXoECAEQAw&biw=1366&bih=583&dpr=1#i
    mgrc=1ANMwvwfON5dAM.
19. https://www.javatpoint.com/applications-of-machine-learning.
20. https://www.geeksforgeeks.org/introduction-to-recurrent-neural-network/.
21. https://www.verywellmind.com/structure-of-a-neuron-2794896.
22. https://www.google.com/search?q=working+of+neural+network&sxs-
    rf=ALiCzsbTBgQBW6ZFpN4aoZto3u2vrVPDtg:1653667233780&source=l-
    nms&tbm=isch&sa=X&ved=2ahUKEwiY7a7AhoD4AhXljuYKHVBmAZYQ_
    AUoAnoECAEQBA&biw=1366&bih=583&dpr=1#imgrc=5FG_zIXEoE9_FM.

5

Intelligent Machining

Jasvinder Singh[1,2]*, Chandan Deep Singh[1] and Dharmpal Deepak[1]

[1]Punjabi University, Patiala, India
[2]Lovely Professional University, Phagwara, Punjab, India

Abstract

Manufacturing industries are constantly forced by market demands to produce low-cost, high-quality products. Market demands are complex and variable, especially in fast-paced industrial environments. Traditional manufacturing procedures are being compelled to create and adapt innovative practices and methods in order to meet increasing market demands. However, as manufacturing technology advances, so does the need to put it into practice in a timely and cost-effective manner. A combination of old production tools and modern technologies, either hardware or software, that transforms traditional manufacturing into intelligent manufacturing is the result of this urgent demand. Intelligent manufacturing systems combine optimization approaches with sensor-based control systems to create complex systems. These intelligent and optimized systems can produce high-quality items at a lesser cost and at a faster rate. New advancements in machining have been possible by latest developments in information and communications technologies, particularly artificial intelligence. Modern machining systems are very sophisticated and automated, but their performance is mostly dependent on the operators' knowledge, which is based on a combination of theory and experience. With the integration of simulation, sensing, modelling, control, and monitoring of the process, intelligent machining attempts to enable intelligent behavior in the machining system. Machine learning and intelligent prediction of machining system behavior are possible due to the recording of process input and output data.

This chapter provides an introduction to intelligent machining. The various components of intelligent machining like Sensors, Machine learning and knowledge discovery, Database knowledge discovery, Programmable Logical controller (PLC), Information integration via knowledge graphs are presented. In starting, requirements for the developments of intelligent machining and subdivision of intelligent machining

---

*Corresponding author: jas.sliet86@gmail.com

Chandan Deep Singh and Harleen Kaur (eds.) *Factories of the Future: Technological Advancements in the Manufacturing Industry*, (103–120) © 2023 Scrivener Publishing LLC

with details of various types of sensors with their features and function are presented; afterwards the role of machine learning and how it is important in intelligent machining, the role and importance of knowledge database are briefly mentioned. Finally, history and use and advantages of PLC, role of intelligent machining for implementation of green manufacturing in manufacturing industries is described.

Keywords: Intelligent machining, sensors, programmable logical controller (PLC), machine learning and green manufacturing

5.1   Introduction

Intelligent Machining is a method of machining that incorporates sensors into the machining process to improve quality, productivity, and decision-making. To fulfil the growing demands for increased product quality, greater product diversity, shorter product lifecycles, and lower costs, manufacturers are increasingly resorting to an intelligent machining paradigm.

Intelligent machining is a manufacturing paradigm in which machine tools can sense their own states as well as the condition of the environment in which they are operating and may initiate, control, and stop machine actions. Intelligent machines, in other words, are self-aware and capable of making judgements and decisions about manufacturing processes. The integration of smart sensors and controllers enables the intelligent machining paradigm.

5.2   Requirements for the Developments of Intelligent Machining

Commercial factors drive the present development of metal cutting machine tools, with the goal of achieving maximum machining efficiency at the minimum production costs. High efficiency is now accomplished through motions along controlled axes, multitasking, easy programmable configuration of the machine tool for diverse machining operations, strong machining operations precision, decreased idle periods, and machining process [1–3]. Many of the criteria are exceedingly difficult to meet, as they necessitate the prevention of various machine tool and machining process disturbances, as well as exceptionally efficient machine tool and process control [4, 5].

Machine tools will be able to self-repair in the future, and machining processes will be able to optimize, supervise, and govern themselves holistically. These requirements include improving the machine tool's autonomy so that it can reliably detect problems in real time while utilizing the inference and analytical skills of an educated operator. The machine-tool should decide on the scope of a repair and how it will be carried out on its own. It is supposed to allow it to successfully increase its precision, eliminate collisions and failure states on its own, and optimize energy usage (compatible with its autonomy). It should be able to assess its own status as well as the status of the processes it is running on a regular basis, and it should have a high level of IT and intelligence built into it. As a result of this, the tooling will be intelligent, relying on virtual models, digitalization, and extremely efficient real-time control [6, 7].

The high degree of organization and increasing autonomy required for the development of such intelligent machine tools is critical. In line with a manufacturing development plan and an adaptive machine tool control optimization strategy, this necessitates improved efficiency and a stronger capacity to cope with disruptions and lessen their consequences. Machine tools' flexibility in terms of technology, as measured by their capacity to adapt to new tasks, must improve [8, 9]. Machine tools' multitasking and hybridity, as well as their reconfigurability, must continue to improve.

Machine tools will very certainly become increasingly reliant on real-time data and virtualization of their framework and processing conditions, as well as their current condition. The plan will be based on integrated concepts that take into account the machine tool's intrinsic interdependencies in terms of statically, thermally, and dynamical properties, as well as operating conditions. For precise control, machine tool system interaction with the external environment will become more essential, and it will be based on current developments in data communication processing, in line with the requirement for intelligent controller of operating characteristics in real time, as well as CAD-CAM and CNC requirements. Control and monitoring will become more integrated in the future. The accuracy of disturbance and error prediction, which is the foundation for active minimization and correction, will considerably improve [10].

The capacity to improve processes in real time will also increase, allowing machine tools and machining processes to have more self-control and self-healing capabilities. Integrated real-time error compensation, which is included into the tool route design and implemented in real time, will become more effective. Machine tools will have quick setup times and perhaps high productivity, which will result in lower product manufacturing costs.

5.3   Components of Intelligent Machining

For Intelligent Machining framework work, we'll need three major components:

- Intelligent sensors for data collection,
- Data acquisition and management system to process and store signals, and
- Machine learning and knowledge discovery component

5.3.1 Intelligent Sensors

NASA first presented the notion of smart sensors while designing a rocket in 1979. Rockets require vast number of sensors to convey data to the ground or spacecraft, such as temperature, position, velocity, and attitude. Even with a powerful computer, processing such enormous amounts of data at the same time is difficult. Furthermore, the computer's volume and weight are limited by the spaceship. When sensors are joined with the microprocessor, it is intended that the sensors will have an information processing function, resulting in the intelligent sensor [11].

A smart sensor is a device of that can detect and sense data from a specific object as well as learn, judge, and analyze signals. It also incorporates a new form of sensor that can communicate and be managed. The intelligent sensor is capable of self-calibration, compensation, and data collection [12]. The precision and resolution of the intelligent sensor, as well as its stability and reliability, as well as its adaptability, are all determined by its competency. When compared to ordinary sensors, smart sensors play fantastic performance-to-price ratio.

The signals from the most recent intelligent sensors are analyzed by the CPU. Microprocessors, which merged the sensor signal conditioning circuit, microelectronic computer memory, and interface circuit on a single chip, resulting in an AI-enabled sensor, were the focus of intelligent sensors in the 1980s [13]. As intelligent measurement technology improved in the 1990s, sensors were able to achieve compactness, incorporation, electronic structure, easiness of use, and ease of operation, but also autonomy, recollection and intellectual functioning, storage systems, inter monitoring, networking and judging functions.

5.3.1.1 Features of Intelligent Sensors

High precision: Intelligent sensors employ a variety of techniques to mitigate the effects of unexpected error, including computer controlled nil modification and zero withdrawal, normal citation genuine computer-controlled contrast total system measurement, complex nonlinear error, a significant amount of constructive processing and interpretation, and a significant amount of real-time data processing.

High maintainability and consistency: The sensor-based module can make up the difference for the impact of zero shift and ability to respond on the working conditions and natural parameters, such as surrounding ambient air and software prepared to significantly contribute instability caused by variance; it can also adjust the variety of observed data, as well as the legitimate idea of self and decision.

High resolution and 'signal-to-noise ratio': The intelligent sensor has storage space, memory, and data processing functions that can eliminate noise from input data and automate the removal of useful data using digital filtration and clustering; artificial neural techniques can eliminate the impact of many parameters under cross-sensitivity circumstances using data fusion.

Strong self-adaptability: The intelligent sensor is capable of assessment, evaluation, and computing. With high/top computers, this can make decisions about each part's power supply and data transfer rate based on the system's operating conditions, allowing the system to run at low power and optimum transmission loss.

Superior efficiency and cost ratio: Intelligent sensor has highest performance, rather than critical phase, the measurement device on its own in all facets of inertial sensor and error checking, meticulously crafted crafts "to obtain, but through the mixture of compute and micro - controller, using electronic circuit technology and inexpensive and strong software, rather than pursuing perfection, the intelligent sensor has superior efficiency, rather than pursuing perfection, because through the combination of compute and microcontroller, using electronic circuit technology and inexpensive and strong software, rather than pursuing perfection, the intelligent sensor has.

5.3.1.2   Functions of Intelligent Sensors

The functioning of an intelligent sensor is approximated by modelling the coordinated activity of a person's sensory systems and brains, as well as long-term technological research and knowledge. It's a fairly self-contained intelligent unit. Its appearance has substantially improved the sensor's performance by software help, reducing the tough requirements of the original hardware performance. The following functions are generally implemented by intelligent sensors:

   i.   We pay attention to the natural occurrences that occur around us. Force, sound, heat, light, electricity and chemistry are all typical messages. Direct and indirect measures are used to measure sensitive elements in most cases.

The advanced function of the intelligent sensor allows it to simultaneously measure a host of diverse and pesticide qualities and offer data that properly depicts the law of materials movement. Temperature, velocity, pressure, and density may all be detected simultaneously using a composite liquid sensor developed by California University. EG & GIC Sensors, an American company, developed a composite mechanical sensor that can simultaneously detect three-dimensional vibrations velocity, velocity, and movement at a spot.

ii. Function-related adaptability Devices and sensors can adjust their qualities remotely in response to changes in a range of circumstances. By accounting for parameter drift caused by ageing components, present technology can prolong the life of a gadget. Simultaneously, it expands its range of operation by autonomously adjusting to changing environmental conditions. The sensor's repeatability and accuracy are improved through adaptive technology. The real correction value for the sampling stage is now the correction plus compensating value, which is no longer an average.

iii. The common sensor's identity, consciousness, and conscience features necessitate regular examination and calibration to ensure that it is in good working condition and accurate. Because an abnormal online measuring sensor is not discovered in a timely way, these fundamental sensor needs must be relocated from the market to the labs or inspecting divisions.

When advanced sensors are used, the situation improves considerably. To begin, self-diagnostic activities on the electrical supply are executed, followed by a diagnostic test to detect the component's failure. Second, based on the length of time it has been utilized, it may be repaired online, and the microprocessor evaluates and proofreads the data contained in E2PROM.

iv. The capacity to store data is often a deciding element in business performance. An intelligent sensor can retain a huge quantity of information that may be accessed at any moment by the user. The device's history might be included in this data.

For example, how long has the sensor been active, and how frequently has the power supply been replaced? It also contains all of the graphs from the sensor, as well as instructions for configuring the sensor and other information. There's also a serial number, a manufacture date, a catalogue, and the final factory test result. The intelligent sensor's storage capacity is the sole constraint to the content. Intelligent sensors provide the capability of digital communication and self-adaptation in addition to the four functions of process information processing, personality, customization, and data storage.

v. Processing of data is a crucial aspect of data processing, and the smart sensor is responsible for this. Not only can the intelligent sensor increase the signal, but it can also digitize it and manipulate it using software. In general, the fundamental sensor cannot produce a linear signal, and process management prioritizes linearity. By using the look-up table mode, the intelligent sensor can linearize the nonlinear signal.

Of course, each sensor's data sheet should be created independently. Another example of intelligent sensor process information processing is filtering the digital signal with a digital filter to eliminate noise or other associated effects. Furthermore, using software to create complex filters is far easier than utilizing discrete electrical circuits. Climate performance is also a significant factor to consider while processing data. The microprocessor can aid in the detection of signals.

The proper coefficient of temperature controller, for example, can be calculated by monitoring the heat of the primary sensing device, allowing for signal temperature control. Nonlinear compensation and other more complicated compensation can also be achieved with software. This is due to the query table's ability to generate practically any curve form. In order to receive their respective data, a variety of different physical qualities must occasionally be measured and processed. The intelligent sensor's emblem controller can easily do operations like as adding, subtracting, multiplying, and dividing many

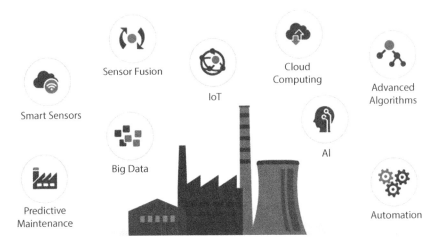

Figure 5.1 How smart sensors make the difference.

signals. The intelligent sensor can play a significant role in the processing of process data. Figure 5.1 representing the role of smart sensors in manufacturing industries.

Furthermore, moving these procedures from the central control room to a place close to the signal generation site is desirable. One reason is because delivering a second signal to the control room is costly, and employing a smart sensor saves money on additional sensors and cables. Another data is recognized at the point of application, reducing the negative impacts of long-distance transmission (such as noise and potential difference) while simultaneously enhancing signal accuracy. The third step is to improve the speed of the control loop by simplifying the software in the primary controller.

vi.   The configuration function is another important component of the intelligent sensor's configuration function.

What is the optimum length of time to magnify the signal? Is the output of the temperature sensor in Celsius or Fahrenheit? Smart sensor users can change their setup at any moment. Only a few of the features available are detection range, programmable test delay, choose groups counting, open/usually shut, and 8/12-bit resolution choice. It's only one of the many permutations available with today's smart sensors. The user's need to design and replace the essential types and numbers of different sensors is substantially reduced thanks to the flexible configuration function. The intelligent sensor's configuration function allows the same device to work in the optimum state and perform different tasks at different times.

vii. As previously stated, digital communication has a purpose. Because intelligent devices may create huge volumes of data and information, regular sensors' single input and output cannot give the essential signal for device data. However, because using a lead line for each piece of data complicates the system, a flexible serial communication system is required.

Point-to-point connections and series networks are common in the process sector. Things are increasingly moving toward a series network. Because it is a digital device with a microcontroller, the smart sensor may easily configure digital communication with the external connection because a serial network can endure significantly more environmental influences (such as electromagnetic interference) than an analogue signal. Serial communication can be successfully handled by matching it to the device so that only required data as an output.

5.3.1.3   Data Acquisition and Management System to Process and Store Signals

Database management starts with the collecting of data and ends with the choice to approve or reject control systems. The steps are as follows, in order:

Data collection: Data management begins with the collection of data processing at the control station; without it, no further actions are possible. In consolidated or dispersed control schemes, the automation in the control center gets data input locally from the controller, or remotely via field devices (RTUs) in network systems. Process data includes the values of analogue variables as well as the status of digital variables in the process.

The input system control station, which acts as a link between both the methods and the implications, makes this possible by collecting physical parameters from processing parameters (values and status), converting it to a digital equivalent without losing information, and sending it to control systems.

- Data transition: The raw data from the processing is turned into facts that may be further processed.
- Data inspection: The next stage is to oversee or watch the process design for any changes in their real levels or conditions, and to generate an audio alarm to alert the operator.
- Visualization of data: The processed data from the management center is shown on the operator's panels and stations so that he or she may have a better grasp of the process's behavior in real time and make necessary decisions. This is the initial part of the human interaction subsystem.
- Data recording and the creation of a history: On-demand, on a regular basis, and in reaction to trends and reports, data-logging stations generate logs. In addition, the automated system records prior processes for future reference. For later off-line study, these records can be preserved on a number of media or platforms.
- Process survey: For a survey, the operator can use their operator stations' screens to display all of the relevant data accessible in the control station (without visiting the process). Whether the process is local or dispersed, the operator can see what is occurring right now (even geographically).
- System investigations: The above-mentioned basic steps are insufficient to generate ideal output, necessitating the execution of specific application programmers to improve the process' performance. The information gathered is utilized to evaluate the process's behavior in order to improve its efficiency.
- Decision-making and data analysis: The machine analyses the given data and makes effective decisions to adjust the process's course, if required, to meet the process's stated goals, all in real time, to manage the process to reach its overall objectives. This stage brings the data management process to a finish. Controlling the processing parameters, as shown in the next section, is how actions are taken.

Process Control

Depending on data analysis and decision-making, this step is utilized to adjust the operation' trajectory if necessary. The automation system executes control systems, or the process of altering the value or state of process variables in a plant, according to a predetermined strategy. The control systems function is usually conducted in real time. Supervisor stations, on the other hand, have the capability of performing human control and actions in order to override the automatic control when necessary (for example, in an emergency).

Receiving control instructions in digital form from control systems or the RTU, converting them into a comparable material form with no data loss, and transferring them to process hardware for control of process parameters is made easier by the output control system sub-system, which serves as the interface between the response framework and the method.

Redesign/Customization

Each process and/or application does not require its own automation system. They are constructed on a unified platform and customized (in hardware and software) to fit the process or application's physical and functional requirements.

5.3.2   Machine Learning and Knowledge Discovery Component

Machine learning is concerned with the creation of computational models that can educate themselves to learn patterns or models from previously collected data and then apply the learnt model to fresh data. Machine Learning systems search through enormous amounts of data for patterns, then use those patterns to make predictions or recommendations based on new data. For example, proposing movie genres that a user might love based on "learned" patterns from previous usage. Machine learning can be used in predictive maintenance and repair applications to anticipate machine service and repair timeframes based on the machine's usage behavior. Machine learning may be used to boost production yields as well. Machine learning techniques are now being used by aerospace, defense, and other high-tech manufacturers to boost production capacity, quality control, and forecasting.

How Machine Learning is Important?

Machine learning is significant because it helps businesses to identify trends in consumer behavior and company processes, as well as assisting in the development of new goods. Many of today's most successful firms, such as Facebook, Google, and Uber, employ machine learning. Machine learning has become a crucial competitive differentiator for many firms.

Types of Machine learning

Machine learning is divided into four categories: supervised, unsupervised, semi-supervised and reinforcement learning. The algorithm that data scientists use is influenced by the sort of data they wish to forecast.

- Supervised Learning: In this sort of machine learning, data analysts provide labelled training data to algorithms and specify the variables they want the program to look for relationships between. The algorithm's input and output also are provided.
- Unsupervised Learning: Algorithms that learn from unlabeled data are used in this sort of machine learning. The application looks for relevant correlations in large data sets. All of the data used to train algorithms is predefined, as are the projections and recommendations they provide.
- Semi-supervised learning: The two prior machine learning algorithms are combined in this method. Despite the fact that data scientists may supply an algorithm with mostly labelled training data, the algorithm is free to explore the data and draw its own judgments about the set.
- Reinforcement Learning: Reinforcement learning is a technique used by data scientists to educate a machine to follow a multi-step process with well-defined rules. Data scientists develop an algorithm to fulfil a job and provide it with positively or negatively feedback as it learns how to do so. However, for the most part, the algorithm chooses which steps to take along the way on its own.

5.3.3   Database Knowledge Discovery

Data mining and knowledge discovery have always been done by hand. As time passed, the volume of data in many systems increased to the point where it could no longer be handled manually. Furthermore, identifying underlying patterns in data is considered crucial for the success of any firm. Artificial intelligence was formed as a result of the development of numerous software tools for detecting hidden information and formulating assumptions.

In the last 10 years, the KDD (knowledge discovery database) method has achieved its peak. Inferential learning, Bayesian statistics, semantic data optimization, knowledge development for intelligent systems, and scientific method are among the numerous methodologies for discovery currently recorded. The end objective is to extract high-level information from low-level information.

Interdisciplinary activities are included in KDD. This process includes data access, as well as scaling algorithms to large data sets and interpreting the findings. The KDD process is supported by data warehousing's data purification and data access operations. Artificial intelligence, which learns empirical principles through testing and observation, aids KDD even more. The data patterns discovered must be repeatable with further data and have some level of confidence. These designs are regarded to be first-of-their-kind. The following are the steps to complete KDD process:

1. Determine the KDD process's aim from the customer's point of view.
2. Identify the application areas in question, as well as the expertise necessary.
3. Choose a data collection or subset of data samples to do discovery on.
4. Clean and pre - process data by figuring out how to manage missing fields and modifying it to meet the needs.
5. Remove any superfluous variables from the data sets to make them easier to comprehend. Examine fascinating features that, depending on the objective or job, can be used to define the data.
6. Use data mining methods to find hidden trends by aligning KDD aims.
7. Investigate hidden patterns using data mining techniques. This method entails determining which models and parameters are appropriate for the full KDD procedure.
8. To uncover relevant patterns, use a specific representational form, such as classification rules or trees, modelling, or grouping.
9. Using the patterns you've discovered, deduce what you need to know.
10. Put the knowledge to good use by incorporating it into another system.
11. It should be documented and reported to everyone who is interested.

5.3.4   Programmable Logical Controller (PLC)

A programmable logic controller (PLC) is an industrial digital computer that monitors the condition of input devices and regulates the state of output devices using customized software.

This form of control system may help almost any manufacturing line, machine activity, or process. The capacity to adjust and repeat the operation or process while recording and transferring crucial data is the most significant advantage of employing a PLC. Modularity is another benefit of a PLC system. To put it differently, you may combine several types of Input and Output devices to meet your individual needs.

History of PLCs

As a relay replacement for GM and Landis, Modicon designed and constructed the first Programmable Logic Controllers. For each new logic configuration, these controllers removed the need for rewiring and extra hardware. The new design improved the functionality of the controls while lowering the amount of cabinet space required for the logic. Dick Morley created the first PLC, model 084, in 1969. The 184, created by Michael Greenberg and released in 1973, was the first commercially successful PLC. Figure 5.2 representing the basic structure of PLC.

The CPU contains an internal software that instructs the PLC on how to do the following tasks:

- Follow the control instructions in the user's program. This application is stored in "nonvolatile" memory, which means it will not be lost if the computer is shut down.
- Other devices, including as I/O devices, programming devices, networks, and even other PLCs, can be interacted with.
- Handle clerical tasks including correspondence, internal diagnostics, and so forth.

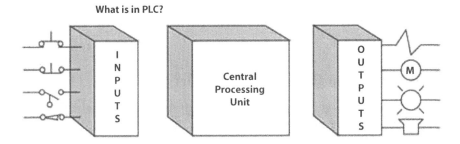

What is in PLC?

Figure 5.2 Structure of PLC.

How does PLC works?

The four basic processes in the functioning of all PLCs are Input Scanner, Program Scan, Production Scan and Maintenance. In a loop, these processes are performed again and again.

The PLC Operation Process Has Four Steps:

1. Input Scan: The condition of all input devices linked to the PLC is detected.
2. Program scan: Implements the logic that the user has created in the application.
3. Production scanning: During scanning, all output devices attached to the PLC are activated or deactivated.
4. Cleaning and maintenance: This process comprises contacts with programming terminals, internal diagnostics, and other similar tasks.

When intelligent machining is paired with the power of the Industrial Internet of Things (IIoTs), connected machines inside a plant may be created, which can then be expanded to connected factories and a fully digitized supply chain. This connected enterprise can play a key role in realizing the Digital Thread by generating, sharing, and utilizing data and information at all stages of the Product (or System) Life Cycle (from conception to disposal) across the supply chain, allowing stakeholders to make better real-time decisions.

5.3.5   Role of Intelligent Machining for Implementation of Green Manufacturing

Machine intelligence, has grown in prominence in the fields of computer science and automation. Automation, information engineering, computer science, mathematics, languages and philosophy all contribute to intelligent machining. As illustrated in Figure 5.3, the difficulties that intelligent machining artificial intelligence (AI) faces have been divided into many sub-problems based on distinct features or skills [7]. Based on their features, the four key problems in green manufacturing may be grouped into three groups:

Knowledge management, dynamic risk evaluation and judgement support, and advance detection are just a few of the techniques that can be used to solve these problems and achieve green manufacturing. Knowledge graphs, Bayesian networks, and deep learning are just a few of

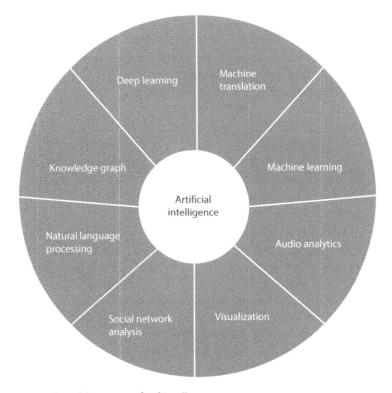

Figure 5.3 Sub-problems in artificial intelligence.

the techniques that can be used to solve these problems and achieve green manufacturing.

5.3.6   Information Integration via Knowledge Graphs

In the field of AI, a knowledge graph is a very well and promising approach for organizing related facts. It's a well-organized semantic network that illustrates the relationships between ideas. Knowledge graphs can also incorporate reasoning and inference abilities based on rules or deep learning algorithms, allowing for the inference of linkages between "entities" inside certain classes. Encyclopedias, social networks, online banking systems, and social security systems are examples of Internet-based applications that leverage knowledge graphs [8, 9].

Chemical engineering, safety training, control systems, automating, and mechanical components in the process sector, in contrary to generic knowledge employed in Web applications, demand a significant lot of specialized

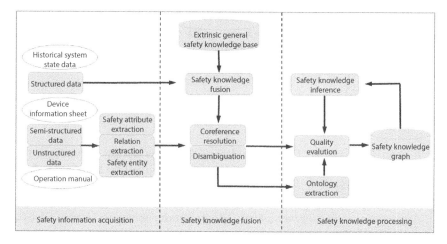

Figure 5.4 Knowledge graph's technical architecture.

in these areas. Implementing knowledge graphs in such a specialized business necessitates not just factual data, but also domain-specific expertise. Modeling data is tough since it requires a thorough grasp of the subject. The construction of a knowledge graph for process safety may be loosely divided into three sections, as illustrated below, based on the fundamental approach for constructing a knowledge graph of any domain (Figure 5.4) [10]. Process safety requires information collecting, knowledge fusion, and knowledge processing.

Sensors are used to keep track on the health of machining processes. Advances in sensor technology and automated data gathering systems have opened up new possibilities for collecting useful measurements in real time and converting them into usable data to help people make better decisions.

Machining, as a robust manufacturing method, can still produce high-quality three-dimensional products from metals, plastics, ceramics, wood, and composites with unequalled capabilities. Researchers have been able to build cost-effective and high-throughput contemporary machining technologies because to advances in computer modelling and optimization approaches.

5.4 Conclusion

When intelligent machining is paired with the power of the Industrial Internet of Things (IoTs), connected machines inside a plant may be created, which can then be expanded to connected factories and a fully digitized supply chain. The linked enterprise can play a critical role in realizing the Digital Thread — During all stages of the Product (or service), generating, sharing, and utilizing data and information.

References

1. Qian, F., Zhong, W., Du, W., Fundamental theories and key technologies for smart and optimal manufacturing in the process industry. *Engineering*, 3, 2, 154–60, 2017.

2. Giffi, C.A., Rodriguez, M.D., Gangula, B., Roth, A.V., Hanley, T., *Global manufacturing competitiveness index*, Deloitte Touche Tohmatsu Limited Global, London, 2016.

3. Consumer & Industrial Products Industry group and the Council on Competitiveness, Deloitte University Press, New York, US, 2016.

4. Williams, E., Environmental effects of information and communications technologies. *Nature*, 479, 7373, 354–8, 2011.

5. Smart Manufacturing Leadership Coalition. *Implementing 21st Century Smart Manufacturing: Workshop Summary Report*, Smart Manufacturing Leadership Coalition, Washington, 2011.

6. Piatetsky-Shapiro, G., Knowledge discovery in personal data versus privacy—A mini-symposium. *IEEE Expert*, 10, 5, 46–47, 1995.

7. Paulheim, H., Knowledge graph refinement: A survey of approaches and evaluation methods. *Semant. Web*, 8, 3, 489–508, 2017.

8. Farber, M., Bartscherer, F., Menne, C., Rettinger, A., Linked data quality of DBpedia, Freebase, Open Cyc, Wiki data, and YAGO. *Semant. Web*, 9, 1, 77–129, 2018.

9. Ehrlinger, L. and Wöß, W., Towards a definition of knowledge graphs, in: *Proceedings of SEMANTICS 2016: Posters and Demos Track*, Leipzig, Germany, 2016, Sep 13–14, 2016.

10. Liu, Q., Li, Y., Duan, H., Liu, Y., Qin, Z., Knowledge graph construction techniques. *J. Comput. Res. Dev.*, 53, 3, 582–600, 2016.

11. Xu, G., Chen, J., Zhou, H., A tool breakage monitoring method for end milling based on the indirect electric data of CNC system. *Int. J. Adv. Manuf. Technol.*, 25, 5, 29–40, 2018.

12. Liu, C. and Jiang, P., A cyber-physical system architecture in shop floor for intelligent manufacturing. *Proc. Manuf.*, 56, 9, 372–377, 2016.

13. Lee, C.H., Yang, M.Y., Oh, C.W. *et al.*, An integrated prediction model including the cutting process for virtual product development of machine tools. *Int. J. Mach. Tools Manuf.*, 90, 3, 29–43, 2015.

6

Advanced Maintenance and Reliability

Davinder Singh* and Talwinder Singh

Department of Mechanical Engineering, Punjabi University, Patiala, Punjab, India

Abstract

Total Productive Management (TPM) is a program that focuses on new ways of repairing equipment throughout the process. This program is designed to improve productivity, decrease production time, and enhance employee satisfaction. The successful implementation of TPM in organizations not only outlines the processes and strategies required for implementation, but also should benefit from the unparalleled results of TPM's core philosophy as a positive view to customers about the company, the quality of its product and reliability. In recent years, many organizations have shown that significant business growth can be achieved through Advanced Maintenance and Reliability (AMR). These ideas and philosophies can be used effectively to improve productivity in an organization, thus leading organizations to a more competitive environment. Therefore, in a highly competitive environment, AMR may appear to be one of the best strategic plans that can lead organizations to measure new levels of success and can really make the difference between success and failure of organizations.

Keywords: Total Productive Management (TPM), Advanced Maintenance and Reliability (AMR), preventive maintenance, predictive maintenance, sustainability

6.1 Introduction

As a pioneer of Total Productive Management (TPM) in the 1940s and 1950s, Japan endeavored to make the most of the ideas and concepts of Preventive Maintenance (PM) in order to develop systems by establishing a Japanese culture of teamwork, cooperation and commitment which later

---

Corresponding author: davinder5206@yahoo.co.in

Chandan Deep Singh and Harleen Kaur (eds.) *Factories of the Future: Technological Advancements in the Manufacturing Industry*, (121–142) © 2023 Scrivener Publishing LLC

known as a total productive maintenance [1]. This innovative approach was rapidly integrated globally and was therefore supported by industry owners [2] in order to reduce maintenance costs and increase system efficiency.

TPM is a program that focuses on new ways of repairing equipment throughout the process. This program is designed to improve productivity, decrease production time, and enhance employee satisfaction [1, 3, 4]. The successful implementation of TPM in organizations not only outlines the processes and strategies required for implementation, but also should benefit from the unparalleled results of TPM's core philosophy as a positive view to customers about the company, the quality of its product and reliability [5].

6.2   Condition-Based Maintenance

For a long time, maintenance was complicated and expensive to support the product life cycle of any particular system. Various methods have been tested to reduce the number of failures or to correct system failures in general [6], by increasing the frequency of maintenance and support functions. The idea of consistent improvement has been tried to overcome the challenges and difficulties it faces in the preparation of the previous generation program. In general, maintenance measures have been divided into two main categories: corrective maintenance (CM) to be performed after detection of a problem while preventive maintenance (PM) is performed prior to diagnosis of the problem; therefore the CM is done randomly while preventive is used over and again. CBM-based maintenance that separates the branch from the PM, should be done after at least one indication that will fail or damage the operation of the equipment. The CBM concept is used in sophisticated mechanical systems that include case reporting and error reporting. It is also used in critical non-technical systems that do not report repetition and error reporting.

Condition-based maintenance is the detection of partial failure or prediction of partial failure. It is a system that recommends acts of maintenance based on condition monitoring information [7]. This information should be strongly correlated with the onset of failure, and a certain amount of limitation should appear to indicate the need for intervention [8].

In the late 1940s, the Rio Grande Railway Steel Company introduced the concept of CBM, and initially, it was called 'predictive maintenance'. The railway company used CBM techniques to check for refrigeration, oil and gas leaks in the engine with variable pressure changes and temperature readings. CBM monitoring strategies have achieved great success in terms of reducing the impact of random failures and set a date to repair leaks or refill the coolant or oil hole. The U.S. military took over the idea early and later used a major maintenance strategy to maintain their military assets [9].

During the 1950s, 1960s, and early 1970s, the concepts and applications of CBM in many industries such as space, automotive, military equipment, and large industries emerged when CBM concepts were adopted and demonstrated many benefits in both efficiency and cost savings [10].

Now large companies and organizations have an interest in CBM ideas and applications including the US Department of Defense (air force, military, navy). Also, other companies such as General Motors, Honeywell, Deglitch, Honda, and General Electric have made significant progress in the field of information technology development in the CBM technology field by enabling bandwidth display, data collection, retrieval, and data analysis and capabilities to support decision-making on large databases of time series data. Targeted data employed in the system or in any system that can provide depth of system performance, system status, the underlying cause of system failure, and the average useful life of the system or sub-system. This serves as a major benefit to the most important systems used in the military, aerospace, and maritime, aviation, automotive, and other industries. These large and important applications have made CBM a major potential source for a company's production line - whether aircraft, vehicles and weapons systems or other products that needs to be repaired from time to time. These industries focus on CBM ideas and maintenance strategies and strategies for designing, planning and CBM agents for players to work in current and future programming engineering.

At a time when the industry is booming, equipment being monitored and supervised by unmanned people as it was before, will increase the need for additional CBM. Unmanned vehicles and systems and systems for robots, wind turbines, oil pump systems, and production systems are just a few examples of systems that can bring many benefits to CBM conservation ideas and strategies. Companies can provide significant improvements in efficiency or revenue if CBM is adopted as a maintenance strategy. This may mean reducing the number of employees, reducing the amount of offers and avoiding costs in the second and third phase failure results, reducing downtime, and other benefits operating in their business environment.

CBM can be done in two ways: online and offline. Online CBM is done when the device is in active mode, while offline CBM is performed when the device is not in working condition. In addition, CBM can be done periodically or continuously. As the name suggests, continuous monitoring is performed continuously and automatically using special measuring devices, such as vibrating and acoustic sensors. Jardine *et al.* [7] identified two key barriers to continuous monitoring.

First, it is expensive because it requires a lot of special equipment. Second, with continuous data collection, the amount of noise increases, which may result in incorrect information. On the other hand, periodic monitoring is performed when continuous monitoring is inappropriate due to economic reasons or factors of equipment failure (e.g., multiple errors). It is done at regular or scheduled times, such as 1 hour or at the end of each shift, with the help of manual processes or portable indicators, such as portable meters, acoustic emission units, and vibration pens. It also includes the use of human sensors to check for mechanical conditions, such as pollution and unusual color. Meanwhile, a major limitation of occasional vigilance is that it may miss certain important information about equipment failure during monitoring intervals.

The second step in CBM is making decision that specifies the tracking action based on the CM output. It may require more detailed analysis in determining the problem and deciding on appropriate remedies, if necessary. This detailed analysis may involve two processes: diagnosis and prognosis. According to Jeong *et al.* [11], diagnosis is the process of finding the source of a problem, while predicting is the process of predicting when a failure may occur [12, 13]. The main purpose of the diagnosis is to provide developers with early warning whenever supervised equipment operates in an emergency. Otherwise, close monitoring and careful maintenance may be required to stabilize the equipment. However, if the instrument is operating in an abnormal state (collapse), it does not mean that the instrument has failed. It may not be used periodically before the failure occurs. Therefore, prognosis is performed. The main purpose of prognosis is to provide additional warning by predicting/estimating when the instrument will fail, so that the instrument can be fully utilized and the decision to make a PM just before the asset fails is determined. From maintenance perspective, prognostic is superior to diagnostic in the sense that prediction can prevent unexpected failures, thus saving additional randomized maintenance costs [7].

6.3   Computerized Maintenance Management Systems (CMMS)

Computers have been used to aid the process of managing repairs since the early 1970s, and by the mid-1980s a large number of repair organizations were using software designed for large computer systems [14]. The software was usually designed in the central computer database where storage and repair information was recorded. The information was then used to generate work schedules and work orders. The producers of the reports allowed the ongoing work to be monitored and the data management information to be produced.

In the late 1980's, a large number of powerful and sophisticated computers were developed. These programs were designed to be human production tools and quickly became commonplace in many organizations. As a result, many repair managers have limited access to complex computer systems for the first time. The result of this increased access was the need for computer-based repair management solutions. To meet these needs, many software houses create building management/maintenance packages. These packages, although similar in concept to the main framework plans of the 1970s, were by nature perfect; using website-related development tools to integrate multiple aspects of the process of remodeling (and management of the building) [15].

Today, one of the most important foundations in the manufacturing industry is undoubtedly the machinery and industrial industries. Increasing productivity and efficiency of production and acquisition of international standards, especially in the domestic and global competitive environment may not be apparent without time management in the use and operation of industrial equipment and production systems, as well as cost management in reducing maintenance costs and rest periods [16].

As a result of technological advances in recent years, the operation, maintenance, and repair of equipment and systems, especially in regard to technological protection and the efficiency of tangible assets, has gradually changed. Organizations are now more aware of the technical efficiency of preventing system failures, as these failures can be a warning of production cuts, loss of customers and a decrease in market share, staff inefficiency etc. According to Tavakkoli-Moghaddam et al. [17], organizations need to develop their understanding of the appropriate design process and implement the system in order to remain competitive in the market. Letters showing organizations should consider that the damage and timing of equipment failures and industrial installations is not a completely preventable issue. However, little attention has been paid to developing planning strategies that can ensure reliability throughout the value chain [18]. It is important because the lack of loyalty to systems puts pressure on organizations to improve overall business performance. Hammer argued that, in order to realize the full potential of the computer system, business processes need to be redesigned primarily to hold the computer, rather than the computer used simply to replace traditional processes. In this way, the computer performs those aspects of the process more efficiently and avoids inherited inefficiencies that have arisen, due to need, in the traditional (manual) process [19].

As discussed in the literature, it is expected that in the near future, CMMS will play a key role in the management of remediation and remediation activities in the most efficient and effective ways [20, 21]. In fact, CMMS plays an important role in solving the problem of large amounts of data collected in an organization, although, due to the complexity of their structures, use of this data is not always possible [22].

CMMS operates increasingly in the control and maintenance of equipment in the service and production of various industries. CMMS principals were first used in hospital rehabilitation programs where mechanical malfunction could have a significant impact on human health [23]. At present, companies consider their value of adjustment management systems as a tool to improve overall performance in their systems [24–26]. The advent of smart devices over the past few years has led to the popularity of these systems in various industries such as health care, automotive and aviation [27–29].

CMMS is a software program designed to assist with the planning, management and administrative functions required for efficient and effective maintenance. Bagadia [30] categorized these functions in producing, organizing, and reporting work orders, improving record tracking and recording segments of tasks. According to Tretten and Karim [31], CMMS is not only a management control tool, but also, it ensures a high level of results through appropriate maintenance activities over time. Apparently, one of the most important benefits that CMMS can provide to the manufacturing industry is that it helps the organization to focus and investigate positive rehabilitation experiences. In general, current CMMSs have been introduced as modular configurations to ensure greater flexibility and optimal compatibility in various production sectors [32, 33].

The use of CMMS can create many benefits for the organization. One of the major benefits is the removal of manual paper operations and the monitoring of system functions that lead to improved productivity [27, 34]. It is noteworthy that the operation of CMMS is the ability to collect and store customized information as required by the user [21]. CMMS does not assist in asset repair decisions, but instead provides personal information to application managers so that they can effectively contribute to service delivery and remain a long-term system.

In general, the benefits of implementing CMMS can be seen as follows [30]:

- Maintain device efficiency by reducing downtime and leading to durable equipment.
- Identify problems that are closer than spotting errors when they occur, leading to minor failures and customer complaints.
- Achieve a high level of organized maintenance services that allow for efficient use of personal resources.
- Influential forces better anticipate asset management and residual purchases to eliminate shortages and reduce existing stocks.
- Maintain high phone performance by minimizing downtime and leading to extended phone life.

6.4   Preventive Maintenance (PM)

Preventive maintenance is an effective maintenance strategy to ensure the regular and effective use of systems and their components. The adjustment based on functioning as a key strategy in high-rise residential buildings is worrying as the consequences of the maintenance resulting from the strategy could not achieve the normal level of performance.

This concept was introduced in 1951, which is a form of physical examination of equipment to prevent mechanical breakdown and extend the service life of the equipment. PM includes repair work done over a period of time or the amount of machine use [35]. During this phase, repair work is established and time-based maintenance activities (TBM) are generally accepted [36]. This type of maintenance depends on the estimated probability that the equipment will break or meet a performance failure over a specified period of time. Protective work performed may include mechanical lubrication, cleaning, replacement, strengthening, and repair. Manufacturing equipment may also be tested for signs of deterioration during repair work [37].

Zhao [38] presented a measure of degradation to represent an incomplete result assuming that the system after the PM action initiates a new failure process. PM policy has also been proposed for a destruction program with an acceptable level of reliability. Bris *et al.* [39] demonstrated the effectiveness of the development method to reduce the cost of PM for a series of compatible systems based on the Birnbaum value-based factor and the use of the Monte Carlo simulation (used by the APLAB planning tool) and genetic algorithm. Badıa *et al.* [40] demonstrate the development of a cost-cutting model for each unit of test period and PM by selecting a unique interval. Gurler and Kaya [41] propose a control policy in which the system is changed when the component falls into a fixed or lower PM position and is placed in a positive, doubtful, appropriate PM and lower class). Motta *et al.* [42] introduced a mathematical analytical and decision-making method that uses loyalty strategies to define the PM's best time for the protection of the power system.

In the literature reported for the period 2001–1997, Salameh and Ghattas [43] determined the level of just-in-time (JIT) baffles by trading the unit cost per unit time and the cost of unit time deficit so that their total amount was minimal. This is shown in the production unit below the normal PM. Tsai *et al.* [44] introduced a system PM with degraded components. Models to reduce age-related degradation of components and genetic algorithm are used to determine the combination of positive functions in each PM. Gupta *et al.* [45, 46] introduces a flexible PM policy of the situation associated with the production environment. It is shown that an extended PM function can reduce the total number of shares expected for a work-in-process (WIP) itself that is without calculating the unplanned minimum. Dohi *et al.* [47] find appropriate PM strategies under the occasional used environment. Lai *et al.* [48] discussed the use of a sequence approach to determine appropriate policy such as when a PM engine action/engine switch should be performed. Ben-Daya and Alghamdi [49] present two consecutive PM models. First, the age reduction of the system is considered to be dependent on the level of PM activity. Second, PM intervals are defined in such a way that the level of risk is the same for all.

Hsu [50] discusses the combined effects of PM and switching policies in a production-like production system with minimal adjustment where it fails and also the modification of JIT production systems for PM disorders. Gopalakrishnan *et al.* [51] introduced a flexible PM system, which maximizes savings of all PM-based products.

6.5   Predictive Maintenance (PdM)

Predictive maintenance includes deciding whether to maintain the system or not according to its nature. In this strategy, maintenance is started by responding to a specific mechanical condition or a malfunction [52]. Diagnostic techniques were used to measure the physical condition of equipment such as temperature, noise, vibration, lubrication and corrosion [53]. When one or more of these indicators reaches a predetermined degradation level, efforts are made to correct the material in order to restore the instrument to its original condition. This means that equipment is removed from work only if there is direct evidence that damage has occurred. Predictive maintenance is based on the same principle as preventive maintenance although it uses a different criterion to determine the need for specific remedial activities. Additional benefit arises from the need for maintenance only when the need is near, not after a certain period of time [54].

It is part of an asset management process that operates on the basis of terms of use. This approach works by creating predictions related to mechanical failure. This method of calculating the working life of an instrument instead works with assumptions related to the failure of the asset. This method works visually for testing, but can also be performed with the help of meters that can measure full-time operation [51]. This approach seems very similar to the prevention method, but is more stable as it reduces the costs associated with testing and maintenance. In addition, the time required for system maintenance is reduced by the location of potential problems and predicting failures.

During the operation of the equipment, the PdM method can monitor the condition and operation of the equipment. The chances of machine failure are reduced by the PdM method of manufacturing in the manufacturing industry. Since the 1990s this PdM method was introduced in the production system due to its practical value. The PdM method has used some advanced technology to predict failure and replacement of machine components such as Internet of Things (IoT), machine learning, artificial intelligence, optimization, etc. [44]. Due to the proper PdM approach, extra maintenance is avoided in planning such as high staff numbers, high component costs, risk of errors and increased waste. Therefore, problem prevention includes necessary adjustments such as critical failures, employee risk and unplanned failure time. For this reason, the IoT method plays an important role in the industry that collects machine data from industries using IoT-based sensors, validated based on previous data and predicted using machine learning-based methods to measure previous failure [48].

6.6   Reliability Centered Maintenance (RCM)

Reliability Centered Maintenance was also established in the 1960s but was originally aimed at aircraft maintenance and used by aircraft, aircraft, and government manufacturers [55]. RCM can be defined as a systematic, logical process to improve or enhance the maintenance of a portable resource in its operational environment in order to recognize its "natural reliability", where "natural reliability" is a level of reliability that can be achieved through efficiency repair program. RCM is the process used to determine the maintenance needs of any material asset in its operational context by identifying asset functions, causes of failure and impacts of failure.

RCM uses seven rational review action philosophies to meet these challenges [56]. Steps include identifying key plant areas, determining key activities and performance levels, determining potential failures, determining potential failures and their consequences, selecting potential and effective maintenance strategies, planning and implementing selected strategies, and developing strategies and programs [57]. The various tools used to influence nutrition development include Failure Mode and Effect Analysis (FMEA), Failure Mode Effect and criticality Analysis (FMECA), Physical Hazard Analysis (PHA), Fault Tree Analysis (FTA), Optimizing Maintenance Function (OMF) and Hazard and Operability (HAZOP) Analysis.

RCM is based on the idea of developing a customized setting in which all possible failures can be accounted for. The strategy works in a way to find the best solution for system failure by working on the type of failure on each machine. RCM is the best remedial strategy that can be used throughout the product life cycle to ensure product development. It can be very effective in product development and product development [43]. Any obstacles to product sales can be managed through an effective repair program.

6.6.1   RCM Principles

A basic way to use reliability-centered maintenance is to improve the overall reliability of the system by creating a system-wide maintenance schedule rather than focusing on individual tasks. Some of the common principles of this method are:

1. Scenario: The RCM approach is based on all operational retention rather than working alone in the operational capacity of the system.
2. Focus: The approach focuses on maintaining the function and not just one component of the system.
3. Reliable: The system holds an account of the operating years of the equipment and the failures it encounters. In short, it is about the failure that can occur with age rather than the rate of failure.
4. Project analysis: The RCM method works by analyzing system structure. It acknowledges the change in reliability to be

a matter of design problems instead of any need for correction. The approach goes through the feedback process which is a good tool for making any changes that have been developed in the design and maintenance of the system [48].

5. Determination method: The RCM method is very important in the security systems of the system maintenance. But overall, it is very concerned about the economics of system maintenance. It works for both economics and asset safety.

6. Unsatisfactory jobs: The RCM proximity model considers failure as unsatisfactory performance. Losses in performance or quality are both considered inappropriate for system performance. The RCM method calculates failures that lead to reduced quality and reduced efficiency.

7. Efficiency: The RCM approach seeks to reduce the likelihood or probability of equipment failure in addition to the costs associated with maintenance at the same time.

8. Adoption: The RCM approach also considers approaches related to maintenance strategies. For example, the RCM has special considerations for initiating a failure or preventative approach as it assesses the need for any intervention function to operate in maintenance.

9. Live system: The RCM method works on data collected over the years. Collected data or work-related history and tools are analyzed in RCM to make any structural changes that may add to system productivity. It is an active maintenance system, and feedback and system history form an important part of the RCM approach [48].

6.7   Condition Monitoring and Residual Life Prediction

Current machine parameters, temperature, vibration, speed etc. they are monitored by a method called monitoring the situation. If an error occurs in the mechanical system then monitoring the situation informs the error diagnosis mode. Therefore, it is a type of PM method that can sense machine parameters using a sensor over complete operating conditions. Sensitive data collected is widely used to predict failure, prevent and predict residual life. The condition monitoring method only ensures the operation of the machine based on failure or low performance. This approach does not interfere with the operation of surveillance equipment. Staff safety is guaranteed and improved equipment reliability [3].

Recently, condition monitoring methods have been used extensively in many manufacturing industries because of their major role in the maintenance process. This method is often used in oil and vibration analysis in many famous industries [58]. The process of this work is divided into three stages, which are described as follows:

- Data collection phase: the state of the data machine is collected by sensors (e.g. IoT-based sensors, edge computing etc.).
- Data processing phase: data collected from phase one is analyzed and processed.
- Decision-making phase: A critical corrective approach is developed based on acquired processing data (e.g., machine learning algorithms, preparation).

After gathering such important information, the adjustment decision model is used for optimal performance. By way of the PM, it is one of the most important units. Equipment failure has prevented the monitoring system and this will avoid failure. However, this approach should be an investment based on short-term value. It is very expensive to monitor the condition of the equipment. Moreover, in the metal cutting system, monitoring the condition of the tool is important [1].

Situation monitoring plays a very important role in the maintenance of industrial property due to its efficiency in detecting potential asset failures, and thus reduces the amount of failure and its consequences. One of the main problems in condition-based maintenance is the prediction of residual health given the partial monitoring parameters to date. This will enable the maintenance decision model to be developed in accordance with other input parameters such as costs or downtime in order to recommend an appropriate adjustment decision. An accurate and accurate forecasting life of an asset is essential in planning and organizing effective and timely cost savings [59].

There are two main methods for predicting residual life, namely the threshold method based on the predetermined preset of the hired monitoring parameter, and the cognitive based method that uses monitored information to determine primary residual life [3].

It is noted, however, that the parameter measured by monitoring the situation is usually first considered by experienced engineers, and then a parameter-based decision is, among other things, made and recommended to repair managers to support their maintenance decisions. This is especially true when monitoring the situation is issued. If the measurement status parameters are accurate and show obvious trends, then both of the methods mentioned earlier can be used depending on the type of precaution [26]. However, it is noted that the measured parameters may contain a lot of noise, therefore, careful analysis and definition of the parameter value is required, forming the basis for expert judgment. As in most cases monitoring situations, specialist judicial information is available, which contains engineer knowledge and expertise, this makes professional judgment information an important source of information to be used in predicting residual health and maintenance decision making [60].

6.8   Sustainability

As in today's dynamic environment, international competition between industries reflects high demand in the manufacturing sector and in society as a whole. Social welfare is affected by the way its economy performs its functioning effectively. Therefore, the state of health depends on the extent to which the economy exploits its resources and meets its aspirations successfully. Achieving sustainable development is important for any society, especially the developing world; however, industrial growth may not be enough to bring about sustainable development. On the other hand, development means an increase in the individual's national income, while development is a means of improving the financial and social status of the poor economies, raising the level of employment, creating improved exploitation of resources and promoting social equality [61].

Many definitions of sustainability have been proposed by various researchers over time. According to the World Commission on Environment and Development, sustainable development is a form of development where the use of assets, investment directives, technical development planning and business transformation are tailored to meet current needs. It is very important to emphasize that countries show different levels of development, from economic growth to economic development and beyond. Sustainable development emerges as an important global vision that we must embrace in order to meet socio-economic, technological and environmental challenges [62].

The manufacturing industry has seen many challenges over the past four decades, including radical changes in innovation, company research and development strategy, export familiarity, flexibility, customer satisfaction and other related issues. These challenges force manufacturing organizations to adopt a new approach to developing new products and using sustainable production tools and techniques. Manufacturing plays an important role in the business of developed countries around the world, but its impact on the environment has become a matter of concern, requiring industries to adopt sustainable production. In other words, it is a matter of doing more with less, i.e. increasing productivity while currently using fewer resources and creating less wasteful waste. Sustainable production also includes aspects of product design; for example, easier disposal for recycling and reduced consumption of hazardous substances [63].

Stability promotes the growth of business processes and market productivity and productivity. Florida [64] proposes a relationship between 'green lean' and environmental sustainability based on literature reviews. Reduced waste and 'less productivity' are directly related to economic and environmental benefits. The system sustainability strategy will be driven by customer needs and environmental conditions. Many researchers have provided a flexible model in which costs are provided by customers and as a manufacturer; we must bring the final product in terms of balance, status and performance. Over the decades many studies have emerged of soft and sustainable production but no clear guidelines for productivity sustainability are available [62].

6.8.1   Role of Sustainability in Manufacturing

Sustainable manufacturing takes into account product development through processes that are non-polluting, energy efficient, environmentally friendly and economical, comprehensive and safe for workers, communities, and consumers. Sustainable production includes the production of 'sustainable' product including production when renewable energy, energy saving and other 'raw' and community-related products are added [65]. Biological monitoring is considered as performance monitoring [66]. Water, energy, and paper consumption are also considered important performance indicators of the environment [67, 68]. It is a way of creating technologies to improve things without the release of greenhouse gases, toxic substances or waste production [69].

A sustainable system is considered closed communication at various levels including financial, environmental and social issues [70]. The U.S. Department of Commerce also cites small negative impacts on the environment, energy, resources, economy and safety of workers, communities covered by a vision of a sustainable production system [71]. Sustainability has complex and varied challenges, some researchers describe sustainability over a long period of time, in terms of a natural perspective while others say it is a powerful environment for sustainability; others see it as a simple process of change to use products [69, 72].

The following factors were emphasized during the study [73]:

- Process evaluation, process, and creativity with metrics and analytical tools
- Technological engagement to reduce impact and produce energy resources
- Green supply chain management in a sustainable environment and production process.

The main function considered in sustainable production is in the process of producing and integrating current recycled products, recycling, reuse and design of new products. Environmental standards enforce the development and implementation of new technologies. Sustainability is a set of skills and concepts that designs a structure and manages its business processes in order to reap the benefits of investing in its large assets that meet the real needs of internal and external stakeholders. The need for sustainability comes from customer needs, height and direction, social behavior, green efforts, use of natural resources and rising costs [73, 74]. All units fit into the product life cycle and create a need for sustainability. Current types of production are designed for production fields, and the system is based on cost, quality, time, and product availability. There is no system model available to meet long-term sustainability. Sustainable metrics help make decisions at all levels of the organization. An important part of sustainable production is the measure of productivity stability by metric selection, the completion of repeated life cycle tests, and the reduction of costs in certain metrics. The need for sustainable production:

Stability is found in the needs of the customers and the main motivations are:

- natural resource development [65],
- global population growth [75],
- global warming [76],
- increased levels of abuse [77],
- the global economy [78].

6.9   Concluding Remarks

The chapter highlights the contributions of various strategies for Advanced Maintenance and Reliability (AMR) in order to gather our benefits of meeting the challenges posed by global competition. AMR has emerged as an important competitive strategy for business organizations in global markets.

In recent years, many organizations have shown that significant business growth can be achieved through AMR. These ideas and philosophies can be used effectively to improve productivity in an organization, thus leading organizations to a more competitive environment. It can be seen as a viable global strategy to provide firms with consistent performance improvements in terms of acquiring critical key skills. Therefore, in a highly competitive environment, AMR may appear to be one of the best strategic plans that can lead organizations to measure new levels of success and can really make the difference between success and failure of organizations.

References

1. Ramesh, V., Sceenivasa Prasad, K., Srinivas, T., Implementation of total productive manufacturing concept with reference to lean manufacturing in a processing industry in Mysore: A practical approach. *ICFAI Univ. J. Oper. Manage.*, 7, 4, 45–57, 2008.

2. Sharma, R. and Singh, J., Impact of implementing Japanese 5S practices on total productive maintenance. *Int. J. Curr. Eng. Technol.*, 5, 2, 818–825, 2015.

3. Fraser, K., Facilities management: The strategic selection of a maintenance system. *J. Facil. Manage.*, 12, 1, 18–37, 2014.

4. Upadhye, N., Deshmukh, S., Garg, S., Key issues for the implementation of lean manufacturing systems. *Glob. Bus. Manage. Res.: Int. J.*, 1, 3, 57–68, 2009.

5. Ahmed, S., Hassan, M., Taha, Z., TPM can go beyond maintenance: Excerpt from a case implementation. *J. Qual. Maint. Eng.*, 11, 1, 19–42, 2005.

6. Wong, E.L., Jefferis, T., Montgomery, N., Proportional hazards modeling of engine', Failures in military vehicles. *J. Qual. Maint. Eng.*, 16, 2, 144–155, 2010.

7. Jardine, A.K., Lin, D., Banjevic, D., A review on machinery diagnostics and prognostics implementing condition-based maintenance. *Mech. Syst. Sig. Process.*, 20, 7, 1483–1510, 2006.

8. Tsang, A.H.C., Condition-based maintenance tools and decision making. *J. Qual. Maint. Eng.*, 1, 3, 3–17, 1995.

9. Prajapati, A., Bechtel, J., Ganesan, S., Condition-based maintenance: A survey. *J. Qual. Maint. Eng.*, 18, 4, 384–400, 2012.

10. Noman, M.A., Nasr, E.S.A., Al-Shayea, A., Kaid, H., Overview of predictive condition-based maintenance research using bibliometric indicators. *J. King Saud Univ.-Eng. Sci.*, 31, 4, 355–367, 2018.

11. Jeong, I.-J., Leon, V.J., Villalobos, J.R., Integrated decision-support system for diagnosis, maintenance planning, and scheduling of manufacturing systems. *Int. J. Prod. Res.*, 45, 2, 267–285, 2007.

12. Lewis, S.A. and Edwards, T.G., Smart sensors and system health management tools for avionics and mechanical systems. *16th Digital Avionics System Conference*, Irvine, C.A, 1997.

13. Vachtsevanos, G., Lewis, F., Roemer, M., Hess, A., Wu, B., *Intelligent fault diagnosis and prognosis for engineering systems*, Wiley, USA, 2006.

14. Pettit, R., Computers aid in the management of housing maintenance. *Proceedings of BMCIS & BRE Seminar, Feedback of Housing Maintenance*, pp. 35–40, 1983.

15. Jones, K. and Burrows, C., Data relationships in the stock condition survey process. *CIB W70 Tokyo Symposium*, vol. 1, pp. 479–86, 1994.

16. Rostamiyan, H., *Total productive maintenance*, Terme Publications, Tehran, 2006.

17. Tavakkoli-Moghaddam, R., Mirzapur, H., Hosseini, A., *Introduction to maintenance planning*, Soft Computing, Mashhad, Sanabad, 2009.

18. Fraser, K., Hvolby, H.H., Tseng, T.L., Maintenance management models: A study of the published literature to identify empirical evidence. A greater practical focus is needed. *Int. J. Qual. Reliab. Manage.*, 32, 6, 635–664, 2015.

19. Ahuja, I.P.S. and Khamba, J.S., Strategies and success factors for overcoming challenges in TPM implementation in Indian manufacturing industry. *J. Qual. Maint. Eng.*, 14, 2, 123–147, 2008.

20. Garg, A. and Deshmukh, S.G., Maintenance management: Literature review and directions. *J. Qual. Maint. Eng.*, 12, 3, 205–238, 2006.

21. Uysal, F. and Tosun, O., Fuzzy TOPSIS-based computerized maintenance management system selection. *J. Manuf. Technol. Manage.*, 23, 2, 212–228, 2012.

22. Litprot, D. and Palarchio, G., Utilizing advanced maintenance practices and information technology to achieve maximum equipment reliability. *Int. J. Qual. Reliab. Manage.*, 17, 144–155, 2000.

23. Saharkhiz, E., Bagherpour, M., Feylizadeh, M.R., Afsari, A., Software performance evaluation of a computerized maintenance management system: A statistical based comparison. *Maint. Reliab.*, 14, 1, 77–83, 2012.

24. Enciso-Medina, J., Multer, W.L., Lamm, F.R., Management, maintenance, and water quality effects on the long-term performance of subsurface drip irrigation systems. *Appl. Eng. Agric.*, 27, 6, 969–978, 2011.

25. Tatila, J., Helkio, P., Holmstrom, J., Exploring the performance effects of performance measurement system use in maintenance process. *J. Qual. Maint. Eng.*, 20, 4, 377–401, 2014.

26. Hooi, L.W. and Leong, T.Y., Total productive maintenance and manufacturing performance improvement. *J. Qual. Maint. Eng.*, 23, 1, 2–21, 2017.

27. Shahin, A. and Ghazifard, A.M., Radio Frequency Identification (RFID): A technology for enhancing Computerized Maintenance System (CMMS). *New Mark. Res. J.*, 3, 5, 13–20, 2013.

28. Tripathi, S. and Tripathi, V.S., An intelligent transportation system using wireless technologies for Indian railways. *JMIR*, 1, 1, 10–17, 2013.

29. Irizarry, J., Gheisari, M., Williams, G., Roper, K., Ambient intelligence environments for accessing building information: A healthcare facility management scenario. *Facilities*, 32, 3/4, 120–138, 2014.

30. Bagadia, K., *Computerized maintenance management systems made easy*, McGraw-Hill, New York, 2006.

31. Tretten, P. and Karim, R., Enhancing the usability of maintenance data management systems. *J. Qual. Maint. Eng.*, 20, 3, 290–303, 2014.

32. Mather, D., *CMMS: A timesaving implementation process*, CRC Press, USA, 2003.

33. Braglia, M., Carmignani, G., Frosolini, M., Grassi, A., AHP-based evaluation of CMMS software. *J. Manuf. Technol. Manage.*, 17, 5, 585–602, 2006.

34. Ramachandra, C.G., Srinivas, T.R., Shruthi, T.S., A study on development and implementation of a computerized maintenance management information system for a process industry. *Int. J. Eng. Innov. Technol.*, 2, 1, 93–98, 2012.

35. Gupta, S.M. and Turki, Y.A.Y., Adapting just-in-time manufacturing systems to preventive maintenance interruptions. *Prod. Plan. Control*, 9, 4, 349–59, 1998.

36. Pai, K.G., Maintenance management. *Maint. J.*, 1, 1, 8–12, October-December 1997.

37. Telang, A.D., Preventive maintenance, in: *Proceedings of the National Conference on Maintenance and Condition Monitoring*, February 14, Government Engineering College, Institution of Engineers, Cochin Local Centre, Thissur, India, pp. 160–73, 1998.

38. Zhao, Y.X., On preventive maintenance policy of a critical reliability level for system subject to degradation. *Reliab. Eng. Syst. Saf.*, 79, 3, 301–8, 2003.

39. Bris, R., Chatelet, E., Yalaoui, F., New method to minimize the preventive maintenance cost of series parallel systems. *Reliab. Eng. Syst. Saf.*, 82, 3, 247–55, 2003.

40. Badía, F.G., Berrade, M.D., Clemente, A.C., Optimal inspection and preventive maintenance of units with revealed and unrevealed failures. *Reliab. Eng. Syst. Saf.*, 78, 2, 157–63, 2002.

41. Gurler, U. and Kaya, A., A maintenance policy for a system with multi-state components: An approximate solution. *Reliab. Eng. Syst. Saf.*, 76, 2, 117–27, 2002.

42. Motta, S., Brandão, D., Colosimo, E.A., Determination of preventive maintenance periodicities of standby devices. *Reliab. Eng. Syst. Saf.*, 76, 2, 149–54, 2002.

43. Salameh, M.K. and Ghattas, R.E., Optimal just-in-time buffer inventory for regular preventive maintenance. *Int. J. Prod. Econ.*, 74, 1, 157–61, 2001.
44. Tsai, Y., Wang, K., Teng, H., Optimizing preventive maintenance for mechanical components using genetic algorithms. *Reliab. Eng. Syst. Saf.*, 74, 1, 89–97, 2001.
45. Gupta, D., Guünalay, Y., Srinivasan, M.M., The relationship between preventive maintenance and manufacturing system performance. *Eur. J. Oper. Res.*, 132, 1, 146–62, 2001a.
46. Gupta, R.C., Sonwalker, J., Chitale, A.K., Overall equipment effectiveness through total productive maintenance. *Prestige J. Manage. Res.*, 5, 1, 61–72, 2001b.
47. Dohi, T., Kaio, N., Osaki, S., Optimal periodic maintenance strategy under an intermittently used environment. *IIE Trans.*, 33, 12, 1037–46, 2001.
48. Lai, K.K., Leung, F.K.N., Tao, B., Wang, S.Y., Practices of preventive maintenance and replacement for engines: A case study. *Eur. J. Oper. Res.*, 124, 294–306, 2000.
49. Ben-Daya, M. and Alghamdi, A.S., On an imperfect preventive maintenance model. *Int. J. Qual. Reliab. Manage.*, 17, 6, 661–70, 2000.
50. Hsu, L., Simultaneous determination of preventive maintenance and replacement policies in a queue-like production system with minimal repair. *Reliab. Eng. Syst. Saf.*, 63, 2, 161–7, 1999.
51. Gopalakrishnan, M., Mohan, S., He, Z., A tabu search heuristic for preventive maintenance scheduling. *Comput. Ind. Eng.*, 40, 1, 149–60, 2001.
52. Vanzile, D. and Otis, I., Measuring and controlling machine performance, in: *Handbook of Industrial Engineering*, G. Salvendy, (Ed.), John Wiley, New York, NY, 1992.
53. Brook, R., Total predictive maintenance cuts plant costs. *Plant Eng.*, 52, 4, 93–5, 1998.
54. Herbaty, F., *Handbook of maintenance management: Cost effective practices*, 2nd ed., Noyes Publications, Park Ridge, NJ, 1990.
55. Dekker, R., Applications of maintenance optimization models: A review and analysis. *Reliab. Eng. Syst. Saf.*, 51, 229–40, 1996.
56. Samanta, B., Sarkar, B., Mukherjee, S.K., Reliability centered maintenance (RCM) strategy for heavy earth moving machinery in coal mine. *Ind. Eng. J.*, 30, 5, 15–20, 2001.
57. Moubray, J., *Reliability centered maintenance*, 2nd ed., vol. 2, pp. 294–306, Industrial Press, New York, NY, 1997.
58. Nash, M. and Poling, S.R., Quality management: Strategic management of lean. *Quality*, 46, 4, 46–49, 2007.
59. Raj, T., Attri, R., Jain, V., Modelling the factor affecting flexibility in FMS. *Int. J. Ind. Syst. Eng.*, 11, 4, 350–374, 2012.
60. Singh, R.K., Garg, S.K., Deshmukh, S.G., Strategic development by SMEs for competitiveness: A review. *Benchmarking: Int. J.*, 15, 5, 525–547, 2008.

61. Salih, T.M., Sustainable economic development and the environment. *Int. J. Soc. Econ.*, 30, 1/2, 153–162, 2003.

62. Attri, R., Grover, S., Dev, N., Kumar, D., Analysis of barriers of total productive maintenance (TPM). *Int. J. Syst. Assur. Eng. Manage.*, 4, 4, 365–377, 2013.

63. Bogue, R., Sustainable manufacturing: A critical discipline for the twenty-first century. *J. Assem. Autom.*, 34, 2, 117–122, 2014.

64. Attri, R., Grover, S., Dev, N., Kumar, D., An ISM approach for modeling the enablers in the implementation of total productive maintenance (TPM). *Int. J. Syst. Assur. Eng. Manage.*, 4, 4, 313–326, 2012.

65. Malian, L. and Walrond, W., *The path to sustainability*, RISI, Boston, USA, 2010, [online] http://www.risiinfo.com/technologyarchives/environment/The-path-to-sustainability.html (accessed May 2016).

66. Haden, S.S.P., Oyler, J.D., Humphreys, J.H., Historical, practical, and theoretical perspectives on green management. *Manage. Decis.*, 47, 7, 1041–1055, 2009.

67. Hajmohammad, S., Vachon, S., Klassen, R.D., Gavronski, I., Lean management and supply management: Their role in green practices and performance. *J. Cleaner Prod.*, 39, 312–320, August 2012.

68. Aguado, S., Alvarez, R., Domingo, R., Model of efficient and sustainable improvements in a lean production system through processes of environmental innovation. *J. Cleaner Prod.*, 47, 141–148, 2013.

69. Brown, A., Amundson, J., Badurdeen, F., Sustainable value stream mapping (Sus-VSM) in different manufacturing system configurations: Application case studies. *J. Cleaner Prod.*, 85, 164–179, June 2014.

70. Kibira, D., Jain, S., Mclean, C.R., *A system dynamics modeling framework for sustainable manufacturing*, The System Dynamics Society, New York, USA, 2009, [online] http://www.systemdynamics.org/conferences/2009/proceed/papers/P1285.pdf (accessed 29 August 2018).

71. US Department of Commerce, *Sustainable Manufacturing Initiative (SMI) and public-private dialogue*, Department of Commerce, Washington, USA, 2010, [online] https://www.oecd.org/sti/ind/45010349.pdf (accessed 22nd November 2018).

72. Hines, P., Found, P., Griffiths, G., Harrison, R., *Staying lean: Thriving, not just surviving*, Lean Enterprise Research Centre, Cardiff University, Cardiff, 2008.

73. Dornfeld, D., *Green issues in manufacturing*, LMAS, Berkeley, USA, 2010, [online] http://lmas.berkeley.edu/public/wp-content/uploads/2010/04/dornfeld-overview-April-2010-1.pdf (accessed 19 August 2011).

74. Bi, Z.M. and Kang, B., Enhancement of adaptability of parallel kinematic machines with an adjustable platform. *ASME J. Manuf. Sci. Eng.*, 132, 6, 1016–1025, 2010.

75. Kabir, M.A., Latif, H.H., Sarker, S., A multi-criteria decision-making model to increase productivity: AHP and fuzzy AHP approach. *Int. J. Intell. Syst. Technol. Appl.*, 12, 3/4, 207–229, 2013.

76. Dornfeld, D.A., Opportunities, and challenges to sustainable manufacturing and CMP. *Mater. Res. Symp. Proc.*, 1157, 919–28, 2009, 03–08, 1157-E03-08.

77. Kaepernick, H. and Kara, S., *Environmentally sustainable manufacturing: A survey on industry practice*, Katholieke Universiteit Leuven, Leuven, Belgium, 2006, [online] http://www.mech.kuleuven.be/lce2006/key5.pdf (accessed 2 August 2018).

78. Westkamper, E., Alting, L., Arndt, G., Life cycle management and assessment: Approaches and visions towards sustainable manufacturing. *CIRP Ann. Manuf. Technol.*, 215, 5, 599–626, 2001.

7

Digital Manufacturing

Jasvinder Singh[1,2]*, Chandan Deep Singh[1] and Dharmpal Deepak[1]

[1]Punjabi University, Patiala, Punjab, India
[2]Lovely Professional University, Phagwara, Punjab, India

Abstract

In today's highly competitive worldwide market, manufacturing industries are constantly looking for innovative ways to reduce lead times and to address customization in new product developments that fulfils client's requirements, including product quality, cost, and aesthetics. Customer demand for newer items with better features is forcing manufacturers to embrace digital technologies. Product and services mass customization is chosen over bulk production. Businesses are under greater pressure to serve a single consumer at a cost comparable to mass production, with the quickest feasible development time and manufacturing by focusing on degrees of customization, quality, and performance, as well as competitive price and value. From the past 50 years, with the advancements made in Information Technology, Material Science, Production Technologies and Supply Chain Strategies manufacturers are well positioned to challenge the traditional methods by which products are designed and manufactured. Digital manufacturing and design encompass visualization, manufacturing simulation, ergonomic and human factor assessments, holistic view of product and process design, product design and development sensitive to process limits and capabilities. This chapter focuses on defining a transformational shift from traditional manufacturing methods to Digital Manufacturing, taking into account factors such as product life cycle and transition, Digital Thread, Digital manufacturing security, Digital manufacturing & CNC machining and additive manufacturing systems.

Keywords: Digital manufacturing & design, digital thread, digital manufacturing security, CNC machining, additive manufacturing

*Corresponding author: jas.sliet86@gmail.com

Chandan Deep Singh and Harleen Kaur (eds.) Factories of the Future: Technological Advancements in the Manufacturing Industry, (143–160) © 2023 Scrivener Publishing LLC

7.1 Introduction

The industrial environment has changed dramatically over the previous few decades as manufacturers attempted to better their performance in a global, linked, and rapidly changing marketplace. Manufacturing firms in the twenty-first century face increasingly frequent and unpredictable market shifts, including the quick introduction of new products and continually changing product demand, as a result of global competition. Traditional dedicated production line approaches, while still in use/practice, are becoming obsolete as market demands change, in an expense and timely way, such as increasing product demand, product modifications, such as the new product in the line, and equipment failure, such as equipment failure. In a highly competitive market and high-volume sector like automotive, the implementation of automation and information systems (ICT) on the shop floor, backed by digital production tools, is important.

Design for manufacturing (DFM), flexible manufacturing (FM), lean manufacturing (LM), and computer-integrated manufacturing (CIM) are examples of manufacturing approaches that demand collaboration across process and product design. The first Industrial Revolution, which began in the mid-1800s, marked the move from manually to automated production, which was aided by the use of water and thermal energy [1]. At the start of the 20th century, the Revolution was welcomed in by the advent of technology that permitted large-scale manufacturing, particularly the mechanical production system [2]. The Third Revolution, often known as the Digital Revolution, occurred between the 1950s and the 1970s, and it symbolized the gains made possible by robotization as mechanical and simple developments gave access to digital equipment. The Fourth Industrial Revolution, sometimes referred to as Industry 4.0, is the next phase in the evolution of industry.

The cyber capabilities coming from developments in computers mixed with physical systems have enabled SMART (Sustainable Manufacturing and Advanced Robotics Technologies) machines that integrate design and manufacture a robotized manufacturing ecosystem [3]. Manufacturing is the transformation of raw resources, segments, or components into finished items that meet the demands of customers. Because it creates a tangible result, manufacturing is the most apparent aspect of the product lifecycle [4].

Industrialized nations throughout the globe have invested heavily in new technology, software, and services in recent years to develop digital manufacturing technology through the use of cyber-physical systems, data analytics, and elevated computing for the best products and systems. Design begins with concept design and finishes with detail design, when finalized part specifications are created, during the phases of the product life cycle. The production of the software from a concept is the next step in the process. This stage comprises establishing the things that must be made, as well as the procedures for creating and assembling them. During the product design and development phase, production prototypes are constructed and tested to ensure that they fulfil performance and cost standards. Long-term manufacturing capacity has also been established [5].

The next stage is the support stage, when the product must continue be supported after it has been made and purchased by the customer. When you provide the customer information to assist them use the product efficiently and maintain its operational performance, you are in the support phase of the life cycle. Disposal and recycling are the last stages of a product's lifetime. The information from the construction stages of the lifespan is used in the disposal phase to determine what a product is composed of and what components of it may be recycled for future purposes [6]. Iterating the design in order to produce a product that meets performance criteria and can be dismantled at the end of its useful life may be crucial at the discard stage. Traditional design and manufacturing processes have been identified as having multiple potential for improvement through time, given to the possibility of many iterations. Several revisions may be necessary during the course of a product's lifespan before the client approves it.

Traditional paper-based methods have a number of drawbacks, such as the fact that information sharing can waste time and cause delays in the development process. When necessary, permissions are requested, also there is a delay in usual procedures throughout as well as between stages of the lifecycle [7]. This can sometimes result in a loss of competitive edge due to being first with the market. Another problem with paper-based processes is the challenge of maintaining a single source of actual product and process specifications, which can limit data analysis and evidence-based decision-making abilities. A new approach called digital manufacturing is the most appropriate way to address all of these difficulties. The notion of digital manufacturing is an integrated suite of technologies that work with product definition data to assist tool design, manufacturing process design, visualization, modelling and simulation, data analytics, and other studies essential to optimize the manufacturing process [8]. Digital manufacturing helps you to communicate information more effectively and commandingly during the design phase.

The phrase "digital thread" refers to a networking structure that enables a linked flow of data and unified view of an investment's data across traditionally fragmented functional viewpoints throughout its existence. It is possible to develop a computer-based digital twin, which is a system of data, models, and analyses that may be utilized in design, manufacture, support and disposal. The notion of a digital thread is critical in smart manufacturing, which strives to coordinate and improve commercial, electronic, and mechanical operations across factories and the whole product value chain. Throughout the product lifetime, the digital thread offers a formal structure for the regulated interchange of authentic technical and as-built data, as well as the capacity to access, combine, convert, and analyze data from diverse systems into actionable information [9].

7.2   Product Life Cycle and Transition

The term "product lifetime" refers to the entire process of developing a product from concept to disposal and recovery. The product lifetime can be viewed from a variety of perspectives. Overall, the item lifecycle is divided into three stages: the beginning of life, which includes design and assembly; the middle of life, which includes use, administration, and upkeep; and finally, the end of life, when these products are re-collected, arranged, dismantled, re-manufactured, reused, and disposed [10]. Product lifecycle data is collected for a variety of reasons by different organizations. Transparency, business assessment, transformation, and estimating are a few of them. Full transparency allows both buyers and producers to make more informed decisions. The arrival of nutrition data on food items and the arrival of natural effects data on products to customers are examples of transparency. To achieve transparency, we must collect and provide information on various aspects of a product or business. Another reason to collect product lifecycle data is to utilize it to evaluate and improve businesses [11].

Continuous improvement necessitates a continuous assessment of the business process from several perspectives. Continuous evaluations necessitate continuous data collecting [12]. Enterprise management can forecast vital information by collecting data. When a company collects market trends in order to forecast future market demands, this is an example of this. Another example is when a company collects data on an industrial machine's energy use in order to predict when it would fail. The product life-cycle is divided into three phases, as we just learned: the beginning, the middle, and the end. During the early stages of life, information is collected using a variety of information management systems, including computer-aided design (CAD) models, computer-aided manufacturing (CAM) systems, and product data management (PDM) systems [13].

After the beginning of the life cycle, the information flow becomes less comprehensive. In fact, for the vast majority of today's consumer products, such as household appliances and automobiles, the information flow is halted once the product is delivered to the customer. This is due to the fact that after things are delivered to clients, ownership of the product is transferred to the user, making it difficult for businesses to track product usage data created by customers. As a result, decision makers in each phase of a product lifecycle make judgments based on incomplete and erroneous product lifecycle data from previous phases, resulting in inefficient decisions in many circumstances. Data from the product lifetime can be divided into two categories: static data and dynamic data [14].

Static data is typically collected at the start of a product's life cycle and rarely changes over time. Bills of materials, material content, take-back information, disassembly instructions, return policies, and recycling information are all examples of static data. Static data is fairly comprehensive and may be acquired using current data management systems. The information created during the usage phase is mostly included in the dynamic information. Designs, natural situations, and modifying actions are examples of progression information [15]. During the product lifecycle, dynamic information is frequently lost and difficult to obtain. The product lifecycle has a wide range of applications. Assume that businesses can track and follow product usage data generated by purchasers and customers [16]. Companies who use product usage data to improve their products and manage relevant operations could achieve significant commercial advantages [17].

Lifecycle acquisition, for example, might be utilized to enable product-related services. Product lifetime data is valuable and may be utilized to help with maintenance, repair, and after-sales services. It's very crucial in the aviation business. Capital equipment and products in the aviation industry have long service lives and complex setups [18]. The company's productivity is derived not only from the sale of capital equipment and flying machines, but also from the maintenance of these assets during a 30-year or longer lifespan. In this vein, maintenance and repair businesses intend to reduce support costs and turnaround times in order to increase revenue.

Data standards are particularly important for creating information or connecting equipment from different manufacturers by establishing a relationship between two separate datasets. They define a standard data format for data sharing and transformation into useable information. STEP and

IGES, for example, are data formats for transferring product design data, such as geometry [19]. MT Connect is a standard that allows data from machining equipment to be captured and shared. To support the product life cycle, Mil Standard 31000A and A SME 1441 specify means to transmit product model information using a technical data package.

7.3   Digital Thread

Starting with the initial concept, products progress through a regular process of plan, design, construct, support, and dispose. During each of these processes, a vast amount of data is generated. Although much of it has not been gathered previously, it has the potential to be translated into incredibly valuable information with a range of implications for the product life cycle. The digital thread has been coined to characterize all of the data created throughout a product's life cycle. From requirement collection to feasibility studies, design, manufacture, testing, and finally sustainment and disposal, The digital thread concept seamlessly connects data across the value chain. This implies that experts may work on product and process definitions at the same time throughout the process, influencing choices throughout the system's or product's life cycle. The digital thread's adoption was gradual at initially, despite its promise, but it has subsequently gained traction. The digital thread technique, like any paradigm shift, has the potential to be disruptive, particularly for businesses without in-house IT capabilities [20]. For example, the information technology basis of the digital thread has traditionally been considered as expense, a cost of doing business that must be lowered.

The IT infrastructure, according to the digital thread paradigm, is an investment that will pay off in enhanced efficacy and responsiveness. A second challenge facing the digital thread is the movement in industrial employment from trained to intelligent or smart workers who combine their talents with practical systems and the ability to interact with data generated throughout the manufacturing process [21]. The production personnel will be more closely linked to the design process as a result of this adjustment. Increasing their ability to influence the design of a product they make. The installation of the digital thread faces a third barrier, which is natural. How are we going to keep the data safe? Physical documents that could be managed and locked up were employed in traditional processes. There's a natural apprehension about putting important intellectual property in digital form, especially given the frequency with which shops seem to disclose intrusions. Data security is an important aspect of the digital thread, and it is embedded in from the beginning [22].

Sharing information and data among the many departments and systems involved in the product life cycle may be extremely beneficial to the overall development process. Faster product launches, improved supplier-to-manufacturer communication, and reduced mistakes, rework, and scrap are just a few examples. Inventory levels in the supply chain are lower, with increased profit margin due to lower unit product costs as well as the expansion into new markets. Techniques for composing and fabricating items are growing at a rapid speed all around us, even in front of our eyes. In this new era of manufacturing, clever sensors and actuators are controlling and monitoring processes that were previously verified and controlled by people [23].

A system's intelligence can be defined as its capacity to attain a goal or maintain desired behavior in the face of unpredictable circumstances. Intelligent machining is a manufacturing criterion in which machine tools can recognize their own states and the environment in which they are functioning, as well as initiate, control, and end machine processes [24]. To put it another way, intelligent robots have a sense of self-worth and can make decisions about assembly procedures. The coordination of keen sensor devices is beneficial to intelligent machining. Advances in sensor technology and automated data systems have opened up new possibilities for collecting valuable measurements in real time and converting them into usable data for making smart decisions [25].

The data aggregator, data post processor, and data buffer components make up the data acquisition and management module. The data aggregator component receives raw data from external sensors that is time synchronized. Because the data collected by various sensors can be in a variety of formats, it must be translated into a common format before it can be processed into information. This interoperability is enabled via data aggregators [26].

Empty connect is an example of an industry standard data aggregating agent. The data processor converts the collected data into formats that machine learning algorithms can use to make better decisions. Data buffer, on the other hand, allows for data buffering so that undesired data can be deleted at this moment. Only relevant data is stored in the data buffer, which can be exploited by machine learning algorithms to provide actionable knowledge. Data buffers serve as a buffer between data processing and machine learning algorithms, allowing for more efficient storage [27]. The data dividend machine learning and knowledge discovery module constructs a data dividend prediction model for the target machine using the processed monitoring data. To monitor and operate the cutting tool, the prediction model is combined with data from previous experiences.

7.4   Digital Manufacturing Security

Security must be considered in all domains, including network, desktop, hardware, and software applications, while using digital manufacturing via the digital thread. Because it involves a variety of valuable intellectual property resources, which we refer to as IP throughout this course [29], security is a vital aspect of the entire digital manufacturing architecture. And it's not to be confused with IP, which stands for Internet Protocols in computer networks. Hardware designs, machine control instructions, payroll operations, and commercial deals are examples of IP resources. Your company could lose a lot of money if your IP information is leaked. As a result, IP must be prioritized while building system security. IP is under threat from a variety of sources, including corporate competitors and state-affiliated entities. The insider threat is one of the most serious hazards to IP. Insider threats are difficult to identify because insiders have legitimate access to sensitive information. As a result, multiple degrees of security are required in DMD security to enable controlled access to information. Only specified authorities inside the organization will be able to access the IP information with this type of access restriction. There are two primary groups of recent advancements in digital manufacturing. The first group's innovations have followed a bottom-up strategy, including manufacturing technologies and developing its concepts inside a bigger framework, such as a digital factory or business. The developments of the second category have followed a model currently, focusing on technology that assists certain aspects of digital manufacturing, such as e-collaboration and simulation.

The digital factory contains concepts, methodologies, and tools for long-term production planning and operations support. It includes operations that are based on digital models that are linked to the product model. On a theoretical level, a number of scholars have contributed to the defining of the digital factory vision, as well as proposals for how this vision may be realized in practice as shown in Figure 7.1. From a technological perspective, guidelines for distributed digital manufacturing have emerged. Recent research has focused on a new generation of 'agent based' decentralized industrial control algorithms. A software agent is a self-directed entity that, first, has its own value systems and a way of communicating with other self-directed objects and, second, acts on its own initiative all of the time. A multi-agent system is made up of a group of agents that are either identical

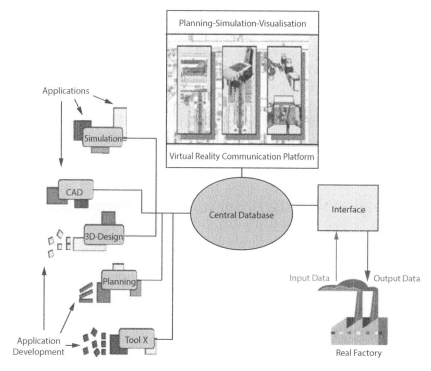

Figure 7.1 The vision of the digital factory.

or complementary and work together. Agent-based systems with real-time and distributed industrial judgement abilities have been reported.

7.5   Role of Digital Manufacturing in Future Factories

Advanced manufacturing, according to the National Institute of Standards and Technology, is an entity that makes extensive use of computer, high precision, and data innovation combined with a superior workforce in a creation framework capable of outfitting a heterogeneous mixture of products in small or large volumes with both the productivity of large-scale manufacturing and the adaptability of custom assembling in order to respond quickly to client demands. Advanced manufacturing, according to Anderson Economy Group, is described as activities that manufacture advanced goods, employ creative manufacturing techniques, and develop new processes and technologies for future manufacturing [28]. Advanced Manufacturing encompasses not only the use of robots and cutting-edge shapes, but also the assembly of inventive goods and the application of cutting-edge standards. Cutting-edge manufacturing components are divided into two categories: tangible and intangible components. Manufacturing processes, industrial control systems, industrial robots, assembly systems, material transportation systems, and storage systems are tangible parts of manufacturing systems. Intangible components of assembly frameworks include all aspects of the firm's planning and administration; examples of cutting-edge technologies include Nano-fabrication, semiconductors, additive manufacturing, and bio-production.

7.6   Digital Manufacturing and CNC Machining

CNC (computer numeric control) machining is a subtractive process in digital manufacturing. Rather than building products layer by layer like additive techniques, CNC machining begins with a strong piece of material and removes what's not required. This item gives a quick review of CNC machining, looks at the operations that fit that category, discusses some design considerations, and looks at what you would have to do once your product is produced.

7.6.1   Introduction to CNC Machining

In CNC machining, computer programming is utilized to control the operation of machining operations. These tools remove excess material from the basic block or billet to create the desired result. Despite the fact that CNC machining is widely used in traditional manufacturing, the digital manufacturing sector has flipped the script. Because of the nonrecurring engineering (NRE) expenses associated with programming CNC machines and designing fixtures to support the machining process, CNC machining formerly had a long payback time and required high volume manufacture to be cost efficient. Digital manufacturing businesses have figured out how to automate both coding and tooling for a wide range of products. As a result of this automation, many digital manufacturers can now make small amounts of CNC machined components for much less money than they could previously.

A collection of computer instructions directs the functioning of the machines in the CNC machining process. These instructions are generated by digital manufacturing's strong computers, which analyze the 3D CAD model you created for your product. The specialized computers that do this analysis also look for flaws that might cause issues with the part's proper assembly. For example, designs are scrutinized to ensure that they do not contain undercuts that are hard to machine. Of course, the evaluation considers a number of other factors as well. Depending on the shape of the intended result, the piece of material may be fixed in place for machining or turned against the tooling.

7.6.2   Equipment's Used in CNC Machining

The CNC machining equipment utilized varies depending on the component being manufactured, However, the tools and the component move relative to each other in every scenario. It's the portion that rotates against the tooling in the case of a lathe. The workpiece is normally fixed in place and the tooling moves with milling machines. Both sorts of processes may be required to fully machine the part in some circumstances. For example, a lathe could be used to turn a cylindrical object that requires a cross-drilled, drilled hole once it has been turned.

Fixturing is a necessary part of CNC machining. It guarantees that the component is always in the proper physical connection with the tooling. Fixturing may be expensive to build, so figure out how much it will cost before you begin. Fixturing isn't charged separately by some digital manufacturing companies, such as Proto Labs, which may save you a lot of money on your components.

7.6.3   Analyzing Digital Manufacturing Design Considerations

CNC machining is a dependable digital production solution, but there are design concerns to keep in mind, as with any manufacturing process. Knowing about these factors can help you plan for the greatest possible parts. When considering how a part could be manufactured utilizing digital manufacturing procedures, It's easy to concentrate on the part's main forms and characteristics. There are a few aspects to bear in mind while using CNC machining for a digital production process.

7.6.4   Finishing of Part After Machining

In manufacturing, secondary procedures are prevalent, especially with metal parts. Heat treatment increases strength and eliminates internal tensions that occur during raw material processing and heavy machining. Carbon steels, such as 1018, can be case hardened by nitriding or carburizing, while 4140 can be quenched and tempered to a hardness of 50 HRc or more. Some 400-series stainless steels, such 174 PH, can be produced quite hard, However, only cold working or drawing through a die can toughen 300-series stainless steels.

Although aluminum and magnesium are soft metals that cannot be hardened, they can be cryogenically strained or aged by mild heating at above baking conditions. Plating is another common post-machining technique. Anodizing is a process that gives aluminum a scratch-resistant surface in practically any color. Nondecorative protection can be achieved with chemical film or chromate. These methods work with magnesium as well, but they require different substances. Because copper and brass discolor when exposed to oxygen, electroless nickel or chrome plating can be applied to protect them. Although steel is routinely coated with black oxide or coated with nickel, cadmium, zinc, and other metals, stainless steel and super alloys do not require this level of protection. Painting is another common alternative, but prior to applying paint, bead blast or other mechanical preparation is required to provide a clean, sludge surface.

7.7 Additive Manufacturing

The additive manufacturing technology creates things by layering material in very thin layers. Extrusion, jetting, fusing, and curing are some of the ways for combining layers of material into solid things. Regardless of the technique used in the process, each layer of material is placed down separately and then bonded to the layers below to make the component.

It is critical to have high-quality equipment when it comes to digital manufacturing. While a low-cost desktop 3D printer would serve for a weekend tinker to make a rough toy army or rudimentary prototype, the technology necessary to produce high-quality models or components for value added services is significantly more complicated, capable, and expensive. The equipment needed for additive manufacturing varies depending on the process, but in order to be effective, it must be able to arrange the material with extreme accuracy. Depending on the procedure, it may also be necessary to fuse or glue the material in place (some processes use materials that immediately connect to the existing material layers). In addition to employing trustworthy additive technologies and materials, producing high-quality prototypes necessitates the use of highly significant computing capability to build precise components. These three advantages are frequently combined by a digital manufacturing organization.

7.7.1   Objective of Additive Manufacturing

Using hundreds of thin layers to construct a part enables for the production of very complex geometries that would otherwise be impossible to machine or mold. Additive manufacturing is very useful for quickly making prototypes, but it may also be used to make small quantities of production parts in some cases. Simply said, designing components without the additional design planning necessary for injection molding or CNC machining allows you to quickly explore design options while avoiding some of the limitations associated with more traditional production techniques.

There are various 3-D printing processes are there like:

- Binder jetting
- Stereolithography
- Fused deposition modelling
- Poly-jet
- Selective laser sintering
- Digital light processing
- Direct metal laser sintering

7.7.2   Design Consideration

A few key design decisions have an impact on how your part can be manufactured. The type of procedure utilized, for example, has a direct impact on factors like the resolution that may be achieved. be a sticking point in terms of perfect part details.

In some cases, it may be necessary to alter your structure to add interior supports. Factors like the part's physical design and material composition dictate whether or not supports are required. If your chosen method necessitates it, your digital manufacturing partner should be able to provide you with support recommendations.

7.8   Role of Digital Manufacturing for Implementation of Green Manufacturing in Future Industries

Manufacturing businesses are increasingly prioritizing environmental effect reduction as a result of greater worldwide attention on global warming and stricter environmental regulations [28]. Manufacturing accounts for more than 35 percent of worldwide CO2 emissions and consumes more than a third of global energy, according to a report issued by the International Energy Agency in 2015 [29]. Meanwhile, through digitalized and intelligent production, industrialization is evolving towards Industry 4.0, which strives to achieve better levels of efficiency and production while using less resources and spending less money.

Industry 4.0 is supported by advances in information and communication technology (ICT) and data storage [3]. The eight primary enabling technologies of Industry 4.0 are Cyber Physical Systems (CPS), Internet of Things (IoT), cloud technology, data analytics, Virtual Reality (VR), intelligent robots, Manufacturing Artificial Intelligence (MAI), and Additive Manufacturing [30–35].

Since the phrase "Industry 4.0" was established at the Hanover Fair in Germany in 2011, it has been a potential strategy for increasing overall operation performance by merging manufacturing and business operations [36]. Industry 4.0, on the other hand, has the potential to bring several prospects for environmental sustainability in addition to economic gains [31]. IoT allows for real-time monitoring and data collection on energy use, allowing manufacturers to optimize and save energy [35, 36]. AM allows for customized design and production, which helps to save resources and reduce waste [37–39]. Because of the decreased transportation costs, CPS provides for a transparent production network with effective communication, resulting in lower emissions [40–42].

However, according to a recent latest research, firms rarely view Industry 4.0 to be good for environmental sustainability, and commercial potential take precedence over environmental and social benefits [43]. There is another research that demonstrates Industry 4.0 has a negative influence on the environment. CPS and Internet integrated circuits and gadgets are widely used and often updated, leading in a considerable volume of e-waste [44]. ICT production and use a rising number of resources, hastening natural resource depletion. The increased need for energy for digitalization and data centers pollutes the environment significantly. As a result, it's necessary to consider both the good and negative effects of digitalization in manufacturing on the environment's long-term viability.

7.9   Conclusion

Digital Manufacturing & Design produces a product at a rate that defies human comprehension. The DMD domain is based on a thorough thread of product, information, and data management, all of which contribute to considerable protected innovation. Industries can move away from paper-based operations and toward digital processes, resulting in increased efficiency. The time it took to develop a product used to be 8–9 months, but now it just takes 20–30 days thanks to digital manufacturing. DMC will become a platform that millions of people will utilize on a daily basis. Users can easily upload their own manufacturing recipes and view their finished products on their own web browsers. To improve efficiency and effectiveness, the digital thread is networked at numerous functions and levels.

DMD lowers supply chain effects by transparently exchanging efficiency data throughout the supply chain, whether it's about people, machines, resources, or procedures. Increasing supply chain trust between layers, removing excess inventory in the supply chain, and realizing more efficient and agile supply chain models. The creation of novel assembling ideal models, such as smart manufacturing, social assembling, and cloud-based manufacturing, has been taken into account by data analytics concepts. The virtual depiction of factories, buildings, resources, machine systems, labor workers, and their capacities, as well as the tighter integration of product and process design via modelling and simulation, are all examples of digital manufacturing. Narrowing the gap among product definition and company's production manufacturing activities within an organization, fully transforming tacit manufacturing knowledge into concrete, and finally, digital awareness, optimizing data management, and emerging standard models are just a few of the top priorities.

References

1. Brugo, T., Palazzetti, R., Ciric-Kostic, S., Yan, X.T., Minak, G., Zucchelli, A., Fracture mechanics of laser sintered cracked polyamide for a new method to induce cracks by additive manufacturing. *Polym. Test.*, 50, 301–308, 2016.
2. Capel, A.J., Edmondson, S., Christie, S.D., Goodridge, R.D., Bibb, R.J., Thurstans, M., Design and additive manufacture for flow chemistry. *Lab. Chip*, 13, 23, 4583–4590, 2013.
3. Choi, S., Jun, C., Zhao, W.B., Noh, S.D., Digital manufacturing in smart manufacturing systems: Contribution, barriers, and future directions, in: *Advances in Production Management Systems: Innovative Production Management Towards Sustainable Growth*, vol. 460, pp. 21–29, 2015.
4. Choi, S., Kim, B., Noh, S., A diagnosis and evaluation method for strategic planning and systematic design of a virtual factory in smart manufacturing systems. *Int. J. Precis. Eng. Manuf.*, 16, 6, 1107–1115, 2015.
5. Cotteleer, M., Holdowsky, J., Mahto, M., (Singer-songwriters), *The 3D opportunity primer*, On: Deloitte University Press, New York, United States, 2014.

6. Coykendall, J., Cotteleer, M., Holdowsky, J., Mahto, M., *3D opportunity in aerospace and defense: Additive manufacturing takes flight. A deloitte series on additive manufacturing*, 2014. 1.Ding, L., Davies, D., McMahon, C.A., The integration of lightweight representation and annotation for collaborative design representation. *Res. Eng. Des.*, Deloitte University Press, New York, United States, 20, 3, 185–200, 2009.

7. Dehoff, R.R., Tallman, C., Duty, C.E., Peter, W.H., Yamamoto, Y., Chen, W., Blue, C.A., Case study: Additive manufacturing of aerospace brackets. *Adv. Mater. Processes*, 171, 3, 19–23, 2013.

8. Delaporte, P. and Alloncle, A.P., INVITED Laser-induced forward transfer: A high resolution additive manufacturing technology. *Opt. Laser Technol.*, 78, 33–41, 2016.

9. Foster, I., Zhao, Y., Raicu, I., Lu, S., Cloud computing and grid computing 360-degree compared, in: *Grid Computing Environments Workshop*, Austin, 2008.

10. Jovanovic, V. and Hartman, N.W., Web-based virtual learning for digital manufacturing fundamentals for automotive workforce training. *Int. J. Contin. Eng. Educ. Life Long Lear. XIV*, 23, 3-4, 300–310, 2013.

11. Kim, H., Lee, J.-K., Park, J.-H., Park, B.-J., Jang, D.-S., Applying digital manufacturing technology to ship production and the maritime environment. *Integr. Manuf. Syst.*, 13, 5, 295–305, 2002.

12. Porter, K., Phipps, J., Szepkouski, A., Abidi, S., (Singer-songwriters), *3D opportunity serves it up: Additive manufacturing and food*, On: Deloitte University Press, New York, United States, 2015.

13. Kitson, P.J., Glatzel, S., Chen, W., Lin, C.G., Song, Y.F., Cronin, L., 3D printing of versatile reaction ware for chemical synthesis. *Nat. Protoc.*, 11, 5, 920–936, 2016.

14. Kurzrock, R. and Stewart, D.J., Click chemistry, 3D-printing, and omics: The future of drug development. *Oncotarget*, 7, 3, 2155–2158, 2016, Retrieved from <Go to ISI>://WOS:000369951800001.

15. Li, B.H., Zhang, L., Wang, S.L., Tao, F., Cao, J.W., Jiang, X.D. *et al.*, Cloud manufacturing: A new service-oriented networked manufacturing model. *Comput. Integr. Manuf. Syst.*, 16, 1, 1–7, 2010.

16. Monostori, L., Váncza, J., Kumara, S.R., Agent-based systems for manufacturing. *CIRP Ann.*, 55, 2, 697–720, 2006.

17. Piller, F., Vossen, A., Ihl, C., From social media to social product development: The impact of social media on co-creation of innovation. *Unternehmung*, 66, 1, 7, 2012.

18. Red, E., French, D., Jensen, G., Walker, S.S., Madsen, P., Emerging design methods and tools in collaborative product development. *J. Comput. Inf. Sci. Eng.*, 13, 3, 031001, 2013.

19. Red, E., Holyoak, V., Jensen, C.G., Marshall, F., Ryskamp, J., Xu, Y., v-CAx: A research agenda for collaborative computer-aided applications. *Comput. Aided Des. Appl.*, 7, 3, 387–404, 2010.

20. Schaefer, D. (Ed.), *Product development in the socio-sphere: Game changing paradigms for 21st century breakthrough product development and innovation*, p. 235, Springer, London, UK, 2014.

21. Shen, W., Hao, Q., Yoon, H.J., Norrie, D.H., Applications of agent-based systems in intelligent manufacturing: An updated review. *Adv. Eng. Inform.*, 20, 4, 415–31, 2006.

22. Sutherland, I.E., Sketch pad a man-machine graphical communication system, in: *Proceedings of the SHARE Design Automation Workshop*, ACM, pp. 6–329, 1964.

23. Tolio, T., Ceglarek, D., ElMaraghy, H.A., Fischer, A., Hu, S.J., Laperrière, L. *et al.*, SPECIES— co-evolution of products, processes and production systems. *CIRP Ann.*, 59, 2, 672–93, 2010.

24. Ulrich, K.T. and Eppinger, S.D., *Product design and development*, vol. 384, McGraw- Hill, New York, 1995.

25. Wu, D., Thames, J.L., Rosen, D.W., Schaefer, D., Enhancing the product realization process with cloud-based design and manufacturing systems. *Trans. ASME J. Comput. Inform. Sci. Eng.*, 13, 4, 041004-1 – 041004-14, 2013, http://dx.doi.org/10.1115/1.4025257. 041004-041004-14.

26. Wang, L., Shen, W., Lang, S., Wise-ShopFloor: A web-based and sensor-driven e-shop floor. *Trans. ASME J. Comput. Inform. Sci. Eng.*, 4, 1, 56–60, 2004.

27. Wu, D., Rosen, D.W., Schaefer, D., Cloud-based design and manufacturing: Status and promise, in: *Cloud-Based Design and Manufacturing: A Service-Oriented Product Development Paradigm for the 21st Century*, D. Schaefer (Ed.), p. 282, Springer, London, UK, 2014.

28. Chen, X., Despeisse, M., Johansson, B., Environmental impact assessment in manufacturing industry: A literature review, in: *Proceedings of the 6th International Eur OMA Sustainable Operations and Supply Chains Forum*, Gothenburg, Sweden, 18–19 March 2019.

29. IEA—International Energy Agency, *World energy outlook*, IEA/Organization for Economic Connection and Development (OCED, Paris, France, 2015.

30. Nascimento, D.L.M., Alencastro, V., Quelhas, O.L.G., Caiado, R.G.G., Garza-Reyes, J.A., Rocha-Lona, L., Tortorella, G., Exploring Industry 4.0 technologies to enable circular economy practices in a manufacturing context: A business model proposal. *J. Manuf. Technol. Manage.*, 30, 607–627, 2019.

31. de Sousa Jabbour, A.B.L., Jabbour, C.J.C., Foropon, C., Godinho Filho, M., When titans meet—Can industry 4.0 revolutionise the environmentally-sustainable manufacturing wave? The role of critical success factors. *Technol. Soc. Change*, 132, 18–25, 2018. [CrossRef].

32. Carvalho, S., Cosgrove, J., Rezende, J., Doyle, F., Machine level energy data analysis—Development and validation of a machine learning based tool. *ECEEE Ind. Summer Study Proc.*, 477–486, 2018.

33. Oláh, J., Aburumman, N., Popp, J., Khan, M.A., Haddad, H., Kitukutha, N., Impact of Industry 4.0 on environmental sustainability. *Sustainability*, 12, 4674, 2020.

34. Merdin, D. and Ersoz, F., Evaluation of the applicability of Industry 4.0 processes in businesses and supply chain applications, in: *Proceedings of the 2019 3rd International Symposium on Multidisciplinary Studies and Innovative Technologies (ISMSIT)*, Ankara, Turkey, 11–13 October 2019, IEEE, Ankara, Turkey, pp. 1–10, 2019, Available online: https://doi.org/10.1109/ISMSIT.2019.8932830 (accessed on 12 September 2020).

35. Ang, J., Goh, C., Saldivar, A., Li, Y., Energy-efficient through-life smart design, manufacturing and operation of ships in an Industry 4.0 environment. *Energies*, 10, 610, 2017.

36. Lins, T. and Rabelo Oliveira, R.A., Energy efficiency in Industry 4.0 using SDN, in: *Proceedings of the 2017 IEEE 15th International Conference on Industrial Informatics (INDIN)*, Emden, Germany, 24–26 July 2017, IEEE, Emden, Germany, pp. 609–614, 2017, Available online: https://doi.org/10.1109/INDIN.2017.8104841 (accessed on 12 September 2020).

37. Mehrpouya, M., Dehghanghadikolaei, A., Fotovvati, B., Vosooghnia, A., Emamian, S.S., Gisario, A., The potential of additive manufacturing in the smart factory Industrial 4.0: A review. *Appl. Sci.*, 9, 3865, 2019.

38. Ford, S. and Despeisse, M., Additive manufacturing and sustainability: An exploratory study of the advantages and challenges. *J. Cleaner Prod.*, 137, 1573–1587, 2016.

39. Song, Z. and Moon, Y., Assessing sustainability benefits of cyber manufacturing systems. *Int. J. Adv. Manuf. Technol.*, 90, 1365–1382, 2017.

40. Thiede, S., Environmental sustainability of cyber physical production systems. *Proc. CIRP*, 69, 644–649, 2018.

41. Brozzi, R., Forti, D., Rauch, E., Matt, D.T., The advantages of Industry 4.0 applications for sustainability: Results from a sample of manufacturing companies. *Sustainability*, 12, 3647, 2020.

42. Berkhout, F. and Hertin, J., De-materialising and re-materialising: Digital technologies and the environment. *Futures*, 36, 903–920, 2004.

43. Kopp, T. and Lange, S., The climate effect of digitalization in production and consumption in OECD countries. *CEUR Workshop Proc.*, 2382, 1–11, 2019.

44. Cosar, M., Carbon footprint in data centre: A case study. *Feb. Fresenius Environ. Bull.*, 28, 2, 600–607, 2019.

8

Artificial Intelligence in Machine Learning

Sikander Singh Cheema*, Er. Lal Chand and Bhagwant Singh

Department of Computer Science and Engineering, Punjabi University, Patiala, Punjab, India

Abstract

In the news and science fiction, artificial intelligence and machine learning have always been hot topics. They've recently received increased media attention due to technological improvements in deep learning and the availability of more consumer-facing apps. Although the phrases machine learning and artificial intelligence are frequently interchanged, there is a distinction. Artificial intelligence is concerned with the concept of "intelligent." Regardless of the underlying approach or algorithm, an A.I. system aims to act as if it has human intelligence. The emphasis in machine learning, on the other hand, is on "learning." Without a person directly programming the knowledge, the system is attempting to learn something from the data. The expert system, for example, was one of the first A.I. achievements. An expert system is one in which the knowledge of a specific topic is put down as rules and directly programmed into the code, allowing the system to respond to questions and complete tasks as if it was a domain expert. This system may appear to have some human intelligence, but it isn't "learning" from data. As a result, this system can be classified as an artificial intelligence (A.I.) system rather than a machine learning system [5]. Artificial intelligence is becoming more widely recognized as a new mobile platform.

Because of the large amount of data produced by gadgets, sensors, and social media users, the computer can learn to recognize patterns and make reasonable predictions. The usage of machine learning and its approaches are discussed in this chapter. Deep learning, which many prominent I.T. providers are using, should also be clarified and debated. Robotics, Artificial Intelligence (A.I.), and Machine Learning (ML) are pushing the limits of what machines can achieve in the software and hardware industries. Automation and innovation have taken them a long way from being a fiction of someone's vision in science-fiction movies and novels to augment human capability (cutting the danger of human errors) in accomplishing jobs faster, more precisely, with more perfection each time.

Corresponding author: sikander@pbi.ac.in

Chandan Deep Singh and Harleen Kaur (eds.) Factories of the Future: Technological Advancements in the Manufacturing Industry, (161–194) © 2023 Scrivener Publishing LLC

Keywords: Artificial intelligence, machine learning, neural networks, classification

8.1   Introduction

Artificial Intelligence (A.I.) is assisting humans in finding solutions to challenging issues. John McCarthy is credited with the invention of artificial intelligence (A.I.). According to him, it's "the science and engineering of developing smart systems and smart computer programs." Computer systems, computer-controlled gadgets, robotic systems, or software emulate human thinking and behavior. Artificial Intelligence is interpreted differently by each researcher, depending on their perspective and point of view. The everyday world is now dominated by technology. Agriculture, medical services, and security services are just a few examples of how technology influences people's life. A user's interaction with a machine is used to collect knowledge about the world. Customization adaptable interfaces and digital aides are suggested in the current Human-Computer Interaction (HCI) investigation to enable quick access to the rising features and services. As machine intelligence grows, a move from HCI to Human-Machine Cooperation (HMC) is required. Human-machine collaboration can be constructed such that machines can examine and accommodate human aims.

The need for people and robots to understand each other's rationale and behaviors is rising. HMC tries to personalize assistance by considering individual user traits, behaviors, and situations to set up HMC so that it can provide the correct knowledge and effectiveness at the right time [18]. HMC has a wide range of applications; yet, there is a knowledge gap in human aspects and artificial intelligence. From HCI task analysis, several approaches and tools are available to design tasks and user interfaces.

In addition, there are several recommendations and norms for HCI in both general and specific application domains. Developing HMC using real-world design techniques is a difficult task. Cognitive engineering is a good fit for this endeavor with its origins in H.F. and A.I.'s leading contributors.

8.2   Case Studies

- **Human-Computer Face Interaction**
  Faces play a crucial role in human-to-human communication [18]. Faces must be found at the heart of human-computer interaction. Intelligent Agents were created for this reason

by Artificial Intelligence (A.I.). They use a machine learning approach to make it easier for people to engage with computers. Experts in machine vision are focusing on building ever-more precise algorithms for recognizing and interpreting face information. Human-Computer Interaction scientists may have brought the prospect of integrating facial data in computer interfaces further than ever before [5]. This chapter's objective is to compile people's choices, observations and questions. HCI, AI, and Cognitive Science can look at computer interfaces and reactions expressed by the human face using all three domains. In recent years, the visual channel has been used to transmit information from the computer to the user. The user's motor channel has been used to send inputs to the computer via the keyboard and pointing devices [18]. Given the rising use of technology that supports the human user in all kinds of chores and activity challenges in affective computing, we reintroduce employing all of our senses into modern computer tools using multimodal devices while creating graphical interfaces. Overall human success in tasks such as logical decision-making, speaking, negotiating, and adapting to uncertain settings is well-documented in neuroscience and psychology. As a result, modern computing platforms are designed to respond to the social demands of their users. We may also be able to gain a better understanding of ourselves by inventing multimodal tools and assisting psychologists in the development and testing of novel ideas of the human cognition and emotion complicated system. This focuses on projects, which involved the result of an automatic facial expression decoder that was primarily used to understand signaled emotions. We offered some of the knowledge from Cognitive Science and Psychology that is related to face expression.

- **Personal Assistant for On-Line Services**
  The On-Line Services Personal Assistant is a person who helps you with your online services [18]. The PALS initiative enhances wireless internet service customer experience. It concentrated on a generic solution, a personal assistant that adjusts the data keynote and navigation support to the user's current context device and interest, and uses context, such as changing the information presentation and navigation support to the user's current context

device and interest. The cognitive engineering method is the first thing that guides the actual realization of a PALS demonstrator. Knowledge and technology were required at various stages. They were created as part of PALS' two discipline-focused research lines, which include enabling the implementation of an effective PALS and expanding the H.F. and A.I. knowledge base, the influence of recognition on mobile user interaction, and A.I. techniques for attuning the exchange to the user's visual attention state. These difficulties were investigated by creating a rule-based in-car system that forecasted the temporary mental load generated by the driving task and adjusted the discourse to avoid overload.

8.3   Advantages of A.I. in ML

AI has many advantages in ML, few has been explained below [16]:

1. A.I. has a favorable impact on every aspect of human life. They range from assisting humans in carrying out routine operations in everyday life to the military, monetary, and health applications. With the undeniable powers of artificial intelligence, it appears to give a paradigm in which humanoid robots and intelligent systems implanted with artificial intelligence can complete and substitute practically all aspects of human life.

2. Artificial intelligence's involvement in medicine is classified into two categories. The first is a conceptual role, while the second is a practical one. This is meant to be a mechanism within a computer for virtual functions [17]. The A.I. component of Machine Learning (also known as Deep Learning) is the identity in practice.

   – The first is machine learning's ability to learn without supervision (unsupervised); in this situation, machine learning's power is quite sophisticated since it can notice patterns created on an event and then realize it on its own. The next sort of machine learning is supervised, which necessitates monitoring and is based on process assessment patterns for classification and classification techniques. The most potent level of machine learning is reinforcement learning, the last category of machine learning [17]. When using reinforcement learning, the machine must 'think' about the optimal course of action when locating an event. Through the experimentation of the activity that he conducted, the device is required to choose which operation will best recompense them.

- As a result, A.I. now provides computerized medical records. In hereditary disorders, a specialized algorithm is employed in electronic medical records to identify the patient's disease history and family history. In addition to hereditary disorders, researchers looked into possible additional dangers when dealing with chronic sickness. Effective collaboration between medical officers and the machine is required for the effective implementation of electronic medical records. The patient's information should be uploaded into the central database in real-time. Furthermore, recognizing that successful communication requires an open-source approach in installing this medical record so that health workers may communicate effectively about the patient's condition [16].

- In addition to virtual applications, A.I. is also used in the physical realm of medicine. The utilization of physical things, electronic medical devices, or the assistance of robots in executing medical chores are all examples of what is meant by physical. Robots as physician assistants are the most likely to be deployed for the last point listed. This is due to the limited mobility that robots can perform; hence Japan makes extensive use of robots in assistance activities. Carebots is the name given to this robot. Carebots are frequently utilized as surgical assistance; in fact, there are now Carebots that can undertake solo procedures [16].

3. Artificial intelligence's favorable impact can also be noticed in the business and economic spheres. Artificial intelligence may be a benefit in disguise for most engineers and CEOs. As is well known, robots capable of doing specified activities instead of human services are the preferred option.

4. Employing robots have several advantages, including more innovative products that are less likely to be tainted by human error. The ability of robots not to feel fatigued is one of its strengths that can be used by individuals in charge of

a company's business flow. Automation and digitization are also thought to increase rather than decrease employment [16]. Furthermore, the existence of digitalization and automation impacts a country's economic growth and income rate, implying a rise in the betterment of the people.

5. AI has many application in agriculture field also like Automation of irrigation system, fertilization automation etc.

8.4 Artificial Intelligence – Basics

Artificial Intelligence (A.I.) is a period used to explain a device's ability to perform cognitive functions similar to those performed by humans, such as perceiving, learning, thinking, and solving problems [1]. The human stage is used as a baseline for A.I. in thinking, speech, and vision.

8.4.1 History of A.I.

1941: The very first computer was an electrical device (technology finally available).
1956: The term "artificial intelligence" was used for the first time.
1960's: A checkers-playing software that could play games with other people.
1980: Systems for quality control.
2000: The first advanced walking robot [1].

8.4.2 Limitations of Human Mind

Object recognition: People are unable to describe how they recognize items adequately.
Face recognition: Cannot be passed on to someone else without explanation.
Naming of colors: Not based on absolute norms, but on learning [1].

8.4.3 Real Artificial Intelligence

General-purpose A.I.:- Just like the robots of technological know-how fiction, is complicated to develop. The human mind seems to have masses of unique and widespread functions, included in a few notable ways that we, in reality, do now no longer recognize at all (yet)
Special-purpose AI:- It is greater doable (nontrivial)

For example, chess/poker gambling programs, logistical planning, automated translation, speech recognition, web search, data mining, clinical diagnostics, and vehicle preservation on the road etc. [2].

8.4.4   Artificial Intelligence Subfields

A.I. is EVERYWHERE [19]

– Machine Translation
– Google Translate
– Spam Filters
– Digital Personal Assistants
– Google Assistant
– Cortana
– Alexa
– Game Players
– DeepBLue
– AplhaGo
– "The Computer" in video games
– Speech Recognition Systems
– Dragon
– Image Recognitions Systems
– Algorithmic Trading Systems
– Black-Scholes Model (Caused crash in 1987)
– Automated Trading Services
– Recommender Systems
– Amazon's Suggestions
– Google Ads
– Autonomous Vehicles
– Self-Driving Cars

8.4.5   The Positives of A.I.

1. Artificial intelligence systems can be beneficial.
2. Digital Personal Assistants help with a variety of daily duties.
3. Spam filters minimize the number of phishing attacks that succeed.
4. Machine translation has aided the flow of information around the globe.
5. Banks utilize artificial intelligence (A.I.) technology to detect fraudulent credit card charges.

8.4.6   Machine Learning

Learning, like intelligence, is hard to define because it encompasses many different processes. "To gain knowledge, insight, competence in, by study, instruction, or experience," according to a dictionary meaning, as well as "improvement of a behavioral propensity by experience." Zoologists and psychologists study animal and human learning. The focus of this chapter is on machine learning. Animal and machine learning share several similarities. Indeed, many machine learning approaches sprang from psychologists' efforts to use computational models to refine their animal and human learning ideas.

It is also possible that the concepts and techniques being studied by machine learning experts shed light on some elements of biological learning. In terms of machines, we may argue that they learn whenever their structure, software, or data is changed to increase their predicted future performance (based on their inputs or in reaction to external information). Some of these modifications, such as adding a record to a database, are more appropriately categorized as part of other specialties and are not certainly properly understood as learning [21]. However, when a speech-recognition system's efficiency increases after hearing multiple examples of a person's speech, we feel justified in claiming that the machine has learned. Improvements in systems are designed to perform tasks connected with artificial intelligence are referred to as machine learning.

Identification, analysis, planning, automatic control, forecasting, and other tasks are among them. The "changes" might either improve existing systems or lead to new ones from the ground up. We show the structure of a typical A.I. "agent" to be more explicit. This agent observes and models its surroundings before calculating appropriate measures, possibly by anticipating their consequences [21]. Changes to any of the elements depicted in the diagram could be considered learning. According to which component is being altered, different learning strategies may be used.

Machine learning is an artificial intelligence field that aims to teach skilled robots to execute complex jobs by studying from enormous volumes of data.

8.4.7   Machine Learning Models

A machine learning model is considered a parameterized function $y = f(x,)$, where x is the input data and is a collection of variables that vary from model to model. The slope and intercept of the line, for illustration, would be the variables in a basic line-fitting model. The variables of a neural network are all of the network's weights. Machine learning's purpose

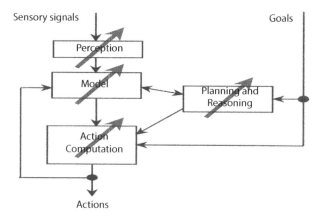

Figure 8.1 Artificial intelligence system [21].

is to discover such that f(x) achieves the desired result y. The issue is that it is frequently high-dimensional and persistent, making the exhaustive search difficult. Instead, we systematically compute the error between the predicted labels and the actual labels using optimal approaches as shown in the Figure 8.1.

Regression models, support vector machines are examples of basic machine learning models. When selecting a model, numerous trade-offs must be considered, including the model's running time, the amount of data required, and the model's effectiveness. Furthermore, models frequently assume things about the information's internal mechanism, which might impact the model's efficiency if the beliefs are incorrect. The naïve Bayes classifier, for example, believes that the input features are unrelated.

8.4.8   Neural Networks

Machine learning's sledgehammers are neural networks. They are appealing, but they necessitate a significant amount of computational power and training data. Because of more advanced machines and the rise of massive data, neural networks have only recently been practical. The firing process of biological neurons, which only fires when the sum of its inputs hits a certain level, is the inspiration for neural networks. Neurons in a neural network, on the other hand, are typically arranged in a succession of layers, with all of the neurons in one layer linked to all of the neurons in the next layer, unlike neurons in the brain, which are distributed messily with links all over the place. A feedforward network, often known as a fully-connected network, is a system like this.

The "firing" is computed by summing the inputs of the neurons to a weighted sum and then performing a nonlinear function to the weighted sum.

For various applications, multiple layouts of neural networks exist. In image processing, a convolutional network calculates activations by sliding a n x n convolutional filter across the 2D picture. This offers the benefit of preserving spatial information, which is essential in picture processing. Recurrent neural networks are used in speech processing tasks like machine translation and natural language understanding because the network preserves an internal state refreshed as the input pattern is processed. This allows the network to keep track of temporal data, such as the order in which words in a phrase are spoken.

8.4.9   Constraints of Machine Learning

Machine learning is suitable for projects with many real datasets, well-defined inputs and outputs, and a measurable way to assess the model's forecast error. The model's efficiency diminishes if any of these requirements are not met. The model learns and is evaluated using a representative sample similar to data seen in the wild. A voice recognition system programmed in American English would undoubtedly struggle when faced with a thick Scottish accent. On hand-drawn drawings, an image classification scheme trained to distinguish things in photographs performs horribly. Acquiring enough data is often one of the most time-consuming and costly aspects of machine learning, as a lousy dataset always results in a flawed model. The architecture of the inputs and outputs is specified in a well-defined task. For example, in a self-driving scenario, the sensor readings are the inputs, and the vehicle's controls are the outputs. Images are the inputs to an image localization algorithm, and the results are bounding boxes around items of interest.

On the other hand, a task like "Decide governmental policy" is not very well defined because the space for viable ideas to enact is both imprecise and infinitely huge. The formalization of generic problems into distinct, well-defined tasks is another hurdle in machine learning. Furthermore, quantifying error is critical because machine learning models are trained by optimizing on some error metric. This mistake, for example, is the difference between the predicted and actual values in regression. Various error measures exist in categorization, but they all penalize selecting the incorrect class. Tasks that need personal assessment or imagination, on the other hand, are frequently unquantifiable. How does one evaluate how "picture-y" the result is when given the job "Draw a picture"? as Figure 8.2 illustrated the relationship between AI, ML, and DL.

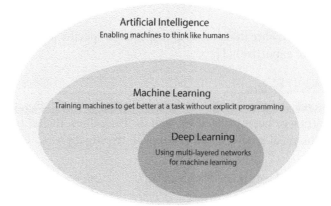

Figure 8.2 Relationship between AI, ML, and DL [3].

8.4.10   Different Kinds of Machine Learning [4]

1. Supervised Learning
2. Unsupervised Learning
3. Reinforcement Learning

1. *Supervised Learning*
   - Someone provides us with examples and the correct responses to those instances.
   - For unseen cases, we must guess the correct answer.
2. *Unsupervised Learning*
   - One sees samples but don't receive any input.
   - One has to look for patterns in the data.
3. *Reinforcement Learning*
   - We take action and get rewards.
   - One will need to figure out how to obtain a lot of money.

8.5   Application of Artificial Intelligence

1. Expert Systems
2. Natural Language Processing (NLP)
3. Speech Recognition
4. Computer Vision
5. Robotics [1]

8.5.1   Expert Systems

The computer software that acts as an expert in a specific topic is an expert system (area of expertise) [1].

Phases in Expert System
Expert systems are increasingly designed to assist rather than replace professionals. They've been utilized in various disciplines, including medical diagnosis, chemical analysis, and geological investigation etc. [1].

8.5.2   Natural Language Processing

The objective of natural language processing (NLP) is to allow humans and computers to interact in a natural (human) language rather than a computer language [1].
 The field of NLP is divided into two categories:

- Natural language understanding.
- Natural language generation.

8.5.3   Speech Recognition

Speech, not reading or writing, is the primary interactive mode of communication utilized by humans. Speech recognition research aims to enable computers to comprehend human speech. They need to be able to hear our voices and understand what we're saying [1]. It improves the objective of NLP by simplifying the process of interactive communication between people and computers.

8.5.4   Computer Vision

People's primary way of experiencing their surroundings is vision; we see more than we hear, feel, smell, or taste in general. Computer vision research aims to provide computers with the same powerful ability to comprehend their surroundings as humans. Here, A.I. aids the computer in learning what it sees through the cameras attached to it [1].

8.5.5   Robotics

A robot is a reprogrammable, multipurpose manipulator designed to move materials, components, tools, or specialized equipment via varied programmed movements to accomplish a range of tasks. An *'intelligent'* robot includes sensory apparatus that allows it to respond to change in its environment [1].

8.6   Neural Networks (N.N.) Basics

A neural network is a computing system modelled after the biological neural networks seen in the human brain. Even though neural networks are not based on a single computer program, they may learn and develop over time.

Neurons are tiny units or nodes that make up a neural network. A link known as a synapse connects these neurons. A neuron can send a signal or information to another neuron nearby via the synapse. The receiving neuron is capable of receiving, processing, and signaling the next call. The procedure is repeated until an output signal is obtained.

8.6.1   Application of Neural Networks

It's important to remember that Neural Networks are a kind of Artificial Intelligence. As a result, the application fields concern systems that attempt to replicate human behavior. Neural networks have a wide range of current applications, including:

> *Computer Vision:* Because no software can be created to make a computer recognize every object, the only method is to employ neural networks, which allow the computer to remember new things on its own over time based on what it already has learned [6].
> *Pattern Recognition/Matching:* This may be used to search a database of pictures for a face that matches a known look. Investigations into criminal cases are an example of this category [6].
> *Natural Language Processing (NLP):* It is a method of teaching and listening to a computer to identify spoken human language [6].

8.6.2   Architecture of Neural Networks

To comprehend artificial neural network architecture, let us understand who built neural network. To describe a neural network made up of many artificial neurons called units organized in layers [6].

Let's have a look at the different layers that an artificial neural network may have.

The Artificial Neural Network is made up of three layers:

Input Layer
It accepts inputs in a variety of formats given by the programmer, as the name implies.
Hidden Layer
It is a concealed layer between the input and output layers. It does all the calculations to uncover hidden features and patterns.
Output Layer
Through the hidden layer, the input goes through a sequence of changes, culminating in output presented using this layer. The weighted total of the inputs is computed by the artificial neural network, which also contains a bias. A transfer function is used to express this calculation [6].

$$\sum_{i=1}^{n} Wi * Xi + b$$

To create the output, the weighted total is provided as an input to an activation function. Activation functions determine whether or not a node should fire. Those who are fired are the only ones who make it to the

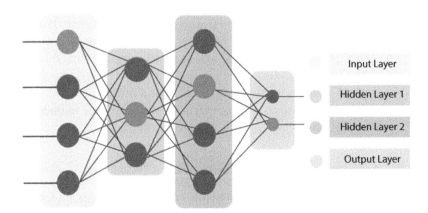

Figure 8.3 Architecture of a neural network [6].

output layer as represented in the Figure 8.3. Many activation functions may be used depending on the type of work [6].

8.6.3   Working of Artificial Neural Networks

The artificial neurons form the nodes of an artificial neural network, which is best described as a weighted directed graph. The directed edges with weights represent the connection between neuron outputs and neuron inputs. The Artificial Neural Network gets the input signal in a pattern and a picture in a vector from an external source. The notations x (n) are used to mathematically allocate these inputs for each n number of inputs.

The weights of every input are then multiplied utilizing every input (those are the information used via the artificial neural networks to resolve a selected problem). These weights usually indicate the strength of the connectivity between neurons inside the artificial neural network in broad terms. Within the processing unit, all of the weighted inputs are summed.

If the weighted total is 0, bias is used to make the output non-zero, or something else is used to scale up the system's response. Discrimination has the same input as weight, and both are equal to one. The total of weighted

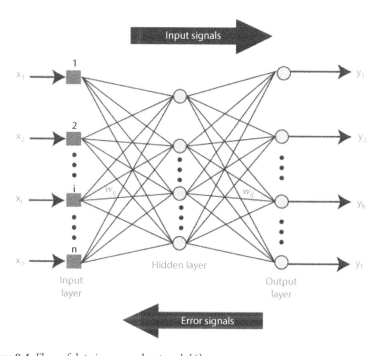

Figure 8.4 Flow of data in a neural network [6].

inputs might be anything from 0 to positive infinity in this case as illustrated in Figure 8.4.

A predetermined maximum value is benchmarked, and the sum of weighted inputs is sent through the activation function to maintain the response within the boundaries of the desired value.

Binary:
The output of a binary activation function is either a one or a 0. A threshold value has been set up to do this. The activation function's final output is one of the net weighted input of neurons is greater than 1; otherwise, it is 0 [6].
Sigmoidal Hyperbolic:
The Sigmoidal Hyperbola function is commonly referred to as a "S" curve. To estimate output from the real net input, the tan hyperbolic function is employed. The following is the definition of the function [6]:

$$F(x) = (1/1 + exp (-????x)) [6]$$

The Steepness parameter is represented by*????* [6]

8.7   Convolution Neural Networks

Computer vision is a deep learning subfield that works with pictures of various sizes. It enables the computer to automatically analyze and interpret the content of a massive number of images.

The convolutional neural network, a derivation of feedforward neural networks, is the basic architecture underpinning computer vision. Picture classification, object identification, neural style transfer, and facial recognition are just a few applications [7].

8.7.1   Working of Convolutional Neural Networks

The improved performance of convolutional neural networks with pictures, voice or audio signal inputs sets them apart from conventional neural networks. They are divided into three sorts of layers:

I.   Convolutional layer
II.   Pooling layer
III.   Fully-connected (F.C.) layer

A convolutional network's initial layer is the convolutional layer. While further convolutional layers or pooling layers can be added after convolutional layers, the fully connected layer is the last. The CNN becomes more sophisticated with each layer, recognizing larger areas of the picture. Earlier layers concentrate on essential elements like colors and borders. As the picture data goes through the CNN layers, it detects more significant components or forms of the item, eventually identifying the desired object [7].

I. Convolutional Layer [20]

- A CNN's main building block is the Convolution layer.
- The parameters are made up of a collection of programmable filters.
- Each filter is tiny in size (width and height) but extends the entire depth of the input volume, for example, $5 \times 5 \times 3$.
- Throughout the forward pass, each filter is slid (convolved) over the width and height of the input volume, and dot products between the filter's values and the input are computed at any place.
- A two-dimensional activation map that shows the filter's reactions in every spatial point.
- The network intuitively learns filters that activate when they see a visual feature.
- Each convolution layer has a set of filters.
  - Each of these generates a two-dimensional activation map, which we may stack along the depth dimension to get the output volume.

The feature detector is a two-dimensional (2-D) weighted array that represents a portion of the picture. The filter size, which can vary in size, is usually a 3x3 matrix, which also defines the size of the receptive field. After that, the filter is applied to a portion of a picture and the dot product between input pixels, the filter is calculated. After that, the dot product sent into an output array. The filter then moves by a stride, and the procedure is repeated until the kernel has swept across the whole picture. A succession of dot products from the input and the filter is a feature map, activation map, or convolved feature [7].

Each output value in the feature map does not have to correspond to each pixel value in the input picture, as seen in the graphic above. It just has to be connected to the receptive field, which is where the filter is applied. Convolutional (and pooling) layers are often called "partially connected" layers since the output array does not have to map directly to

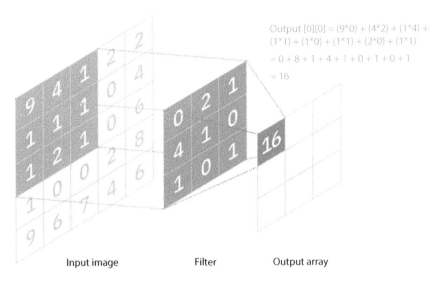

Output [0][0] = (9*0) + (4*2) + (1*4) +
(1*1) + (1*0) + (1*1) + (2*0) + (1*1)

= 0 + 8 + 1 + 4 + 1 + 0 + 1 + 0 + 1

= 16

Input image   Filter   Output array

Figure 8.5 Feature detector image [7].

each input value as shown in Figure 8.5. This trait, however, can also be referred to as local connection [7].

The depth of the output is affected by the number of filters used. Three unique filters, for example, would result in three different feature maps, resulting in a depth of three. The kernel's stride is the number of pixels that travels across the input matrix. Although stride values of two or more are uncommon, a longer stride results in a lesser output. When the filters don't fit the input image, zero-padding is generally employed. All members outside the input matrix are set to zero, resulting in a more significant or equal-sized output [7].

Padding comes in three varieties:

> *Valid padding:* This is sometimes referred to as "no pad-ding." If the dimensions do not align, the final convolution is omitted.
> *Same padding:* This padding guarantees that the output layer matches the input layer in size.
> *Full padding:* By padding the input's border with zeros, this form of padding enhances the output's size.

A CNN performs as a Rectified Linear Unit (ReLU) modification to the feature map after each convolution operation, imparting nonlinearity to the model.

II. Pooling Layer

- Adding the layer of pooling
- It lowers the representation's spatial size
- It controls overfitting by reducing the number of parameters and computations in the network.
- Using the MAX operation, the Pooling Layer operates independently on each depth slice of the input and resizes it spatially.
- The most typical configuration is a pooling layer with 2x2 filters applied with a stride of 2 — this down samples every depth slice in the input by 2 in both width and height.
- The MAX operation would require a total of four numbers (tiny 2x2 region in some depth slice).
- The depth dimension is unaffected [20].

Max pooling:
The filter picks the pixel with the highest value to transmit to the output array as it travels across the input. In comparison to average pooling, this method is utilized more frequently.

Average pooling:
The average value inside the receptive field is calculated as the filter travels over the input and is sent to the output array.

While the pooling layer loses much information, it does provide a few advantages for CNN. They assist in reducing complexity, increasing efficiency, and reducing the danger of overfitting.

III. Fully-Connected Layer

- Neurons in a fully connected layer have complete connections to all activations in the previous layer.
- Their activations can be computed with a matrix multiplication followed by a bias offset [20].

Converting F.C. Layers to CONV. Layers

- The primary distinction between the F.C. and CONV. layers is that the CONV layer's neurons are only connected to a small portion of the input, and many of the neurons in a CONV volume share parameters.

- The neurons in both levels, however, continue to compute dot products. Therefore their functional form is the same.

Converting F.C. Layers to CONV. Layers

- There is an F.C. layer for each CONV layer that implements the same forward function.
- The weight matrix would be a big matrix with primarily zero weights except for a few blocks (owing to local connectedness) where many weights are equal (due to parameter sharing).
- Any F.C. layer, on the other hand, can be transformed to a CONV layer.
- Take, for example, an F.C. layer with K = 4096 that looks at a 757512 input volume.
- If F = 7, P = 0, S = 1, K = 4096 is utilized, F = 7, P = 0, S = 1, K = 4096 can be expressed as a CONV layer.
- In other words, we are making the filter the same size as the input volume. Therefore the output may be 114096 because only a single depth column "fits" across the input volume, yielding the same result as the F.C. layer.

Case Studies

- LeNet. Yann LeCun developed the first successful applications of Convolutional Networks in the 1990s. It was used to read zip codes, digits, and more.
- AlexNet. Popularized Convolutional Networks in Computer Vision, developed by Alex Krizhevsky, Ilya Sutskever and Geoff Hinton.
- The AlexNet was submitted to the ImageNet ILSVRC challenge in 2012 and significantly outperformed the second runner-up (top 5 error of 16% compared to runner-up with 26% error). The network had a very similar architecture to LeNet, but was deeper, bigger, and featured Convolutional Layers stacked on top of each other.
- AlexNet by tweaking the architecture hyperparameters – expanding the size of the middle convolutional layers and making the stride and filter size on the first layer smaller.

8.7.2   Overview of CNN

A convolutional neural network is, in general, a collection of the actions listed below as shown in Figure 8.6 [8]:

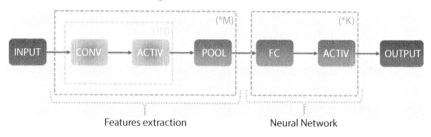

Figure 8.6 Layout of CNN [8].

We use pooling and repeat this procedure a specific number of times after repeating a series of convolutions followed by activation functions. These processes make it possible to extract features based on the picture fed to a neural network; the ultimately linked layers, followed by activation functions regularly, are detailed.

In three dimensions, a convolutional neural network looks like this as shown in Figure 8.7:

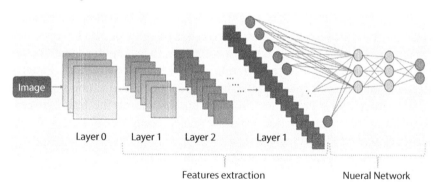

Figure 8.7 3-D layout of CNN [8].

8.7.3   Working of CNN

For two primary reasons, convolutional neural networks provide state-of-the-art image processing results [8]:

> *Parameter sharing*
> A feature detector in the convolutional layer that is beneficial in one area of the image may also be helpful in another.

Sparsity of connections
Each output value in each layer is based on a small number
of inputs.

8.8   Image Classification

Classification between the objects is an easy task for machines. The rise of high-capacity computers, the availability of high quality and low-priced video cameras and the increasing need for automatic video analysis have generated an interest in the object classification algorithm [13].

The classification system consists of a database containing predefined patterns that compare with detected objects to classify into the proper category.

8.8.1   Concept of Image Classification

The technique of mapping numbers to symbols is known as image categorization [9].

f (.) – is a function assigning a pixel vector x to a single class in the set of classes Δ.

The link between the data and the classes into which they are categorized must be adequately understood to classify a piece of data into distinct groups or categories. To do so with a computer, the computer must first be trained. The success of categorization relies heavily on training. Pattern Recognition research spawned the development of classification methods.

The process of a computer program understanding the link between the data and the information classes is called computer classification of remotely sensed pictures.

Crucial Aspects of Proper Categorization

- Learning techniques
- Feature sets

8.8.2   Type of Learning

1. Supervised Learning
   - The learning process is intended to create a link between one set of variables (data) and another set of variables (information) (information classes).
   - The presence of a teacher aids the learning process.

2. Unsupervised learning
- Learning takes place without the presence of a teacher.
- Exploration of the data space to find the scientific law that governs data distribution [9].

8.8.3   Features of Image Classification

Features are properties of data components that are used to categorize them into different groups. A medical diagnosis can be made using temperature, blood pressure, lipid profile, blood sugar, and other pathological data. The characteristics might be qualitative (high, moderate, or low) or quantitative (high, moderate, or low).

The classification may be the presence of heart disease (positive) or absence of heart disease (negative) [9].

8.8.4   Examples of Image Classification

a) Multiple Class Case
   Character or digit recognition from scanned text bitmaps
b) Two Class Case
   In a scanned document, distinguishing between text and visuals [9].

8.9   Text Classification

The practice of classifying text into structured groupings is known as text classification, sometimes known as text tagging or text categorization. Text classifiers use Natural Language Processing (NLP) to evaluate text automatically and assign predefined tags or categories depending on its content [10].

8.9.1   Text Classification Examples

Text categorization is becoming an increasingly essential element of organizations since it enables easy data analysis and business process automation. The following are some of the most famous examples and use cases for automated text classification as shown in Figure 8.8 [12]:

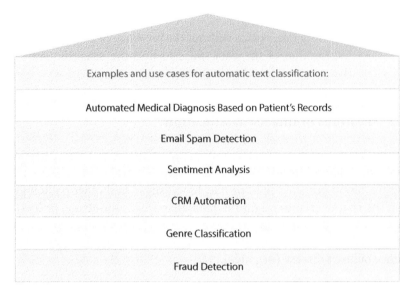

Figure 8.8 Layout of text classification [10].

Sentiment Analysis:
The process of determining if a text is speaking positively or negatively on a topic (e.g. for brand monitoring purposes) [12].

Topic Detection:
The task of deciding the theme or subject of a piece of writing (for example, when analyzing customer feedback, determine whether a product evaluation is about Ease of Use, Customer Support, or Pricing.) [12].

Language Detection:
A method for determining a text's language (For example, knowing whether an incoming support ticket is written in English or Spanish allows keys to be immediately sent to the relevant team) [12].

8.9.2   Phases of Text Classification

Consider a firm that wants to evaluate consumer interest in several product categories. Assume they wish to examine their chat support data to understand better their customers' comments and interest in their various products as shown in Figure 8.9 [10]:

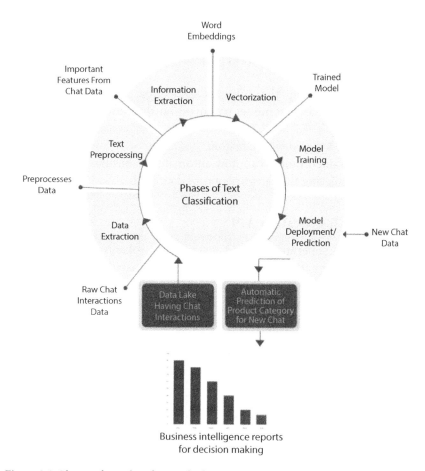

Figure 8.9 Phases of text classification [10].

1) *Data Extraction:*
   Extract valuable data from web pages, data lakes, databases, and other data sources [10].
2) *Text Pre-Processing:*
   We are extracting/retrieving important information/insights from the corpus via parsing, cleaning, and extracting/recovering text [10].
3) *Information Extraction:*
   We employ NLP techniques such as dependency parsing and named-entity identification and feature engineering, and dimensionality reduction to evaluate textual data [10].

4) *Vectorization:*
    For further processing, map words or sentences to a matching vector of real numbers [10].

5) *Model Training:*
    Depending on the business challenge, an appropriate machine learning model is trained on the above word vectors [10].

6) *Model Deployment Prediction:*
    This trained model can now automate the business process by predicting the category of new data [10].

8.9.3   Text Classification API

All preparation operations (text extraction, tokenization, stopword removal, and lemmatization) necessary for automated classification are handled by the Text Classification API.

This API may be used to handle several text categorization situations, including:

Spam filtering (HAM, SPAM) or fundamental emotion analysis are examples of binary classification (POSITIVE, NEGATIVE) [11].

Multiple varieties, such as choosing one category from a list of several – movie genre categorization (thriller, terror, romantic, etc.); multilabel categorization is the process of assigning all applicable types to a single document [11].

Categorization of complex taxonomies – giving categories to a multilayer taxonomy [11].

8.10   Recurrent Neural Network

A recurrent neural network (RNN) is an artificial neural network that works with time series or sequential data. These deep learning algorithms are widely employed for ordinal or temporal issues like language translation, natural language processing (NLP), speech recognition, and picture captioning. They're utilized in popular apps like Siri, voice search, and Google Translate. Recurrent neural networks, like feedforward and convolutional neural networks (CNNs), learn from training input [14].

They are characterized by their "memory," which allows them to alter current input and output by using knowledge from previous intakes. While typical deep neural networks believe that inputs and outputs are independent, recurrent neural networks' production relies on the sequence's primary

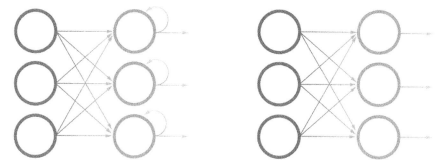

Figure 8.10 Recurrent neural network vs. feed forward neural network [14].

components. While future occurrences may help define a sequence's outcome, unidirectional recurrent neural networks cannot account for them in their predictions as shown in Figure 8.10 [14].

Recurrent and Feedforward Neural Networks (on the left and right, respectively) are contrasted (on the right).

8.10.1   Type of Recurrent Neural Network

Feedforward networks map one input to one output, and while recurrent neural networks are shown in this fashion in the diagrams above, they do not have this limitation. Instead, the length of their inputs and outputs may vary, and different types of RNNs are employed for diverse applications, such as music production, sentiment categorization, and machine translation. The following diagrams are commonly used to represent different types of RNNs:

8.11   Building Recurrent Neural Network

- Feedback loops in networks (Recurrent edges)
- At this moment, the output step is determined by the current input and the initial state (via recurrent edges) [15].

Training RNNs

- Backpropagation Through Time (BPTT) there are four type of Training RNNs as shown in Figure 8.11 for one to one,

One-to-one:

Figure 8.11 One-to-one [14].

One-to-many:

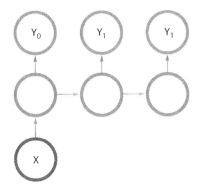

Figure 8.12 One-to-many [14].

Many-to-one:

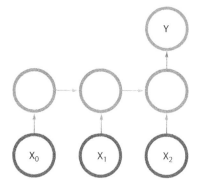

Figure 8.13 Many-to-one [14].

Many-to-many:

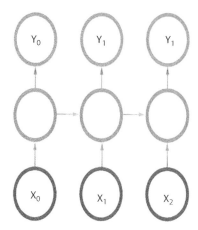

Figure 8.14 Many-to-many [14].

Figure 8.12 for one to many, Figure 8.13 Many to One and Figure 8.14 for Many to Many.

- Regular (feedforward) Backdrop Applied to RNN Unfolded in Time
- Truncated BPTT Approximation

Problem: Backpropagation fails to capture long-term dependencies owing to vanishing/exploding gradients.

8.12 Long Short Term Memory Networks (LSTMs)

A form of RNN design that solves the vanishing/exploding gradient problem and allows long-term dependencies to be learned. It has recently come to prominence with state-of-the-art speech recognition, language modelling, translation, and picture captioning as shown in Figure 8.15 and Figure 8.16 where as Figure 8.17 illustrated the flow of RNN. [15].

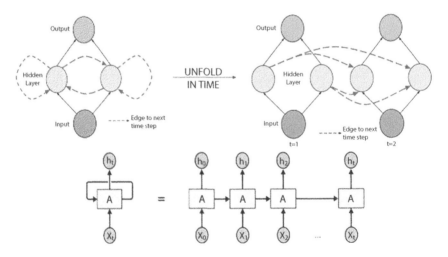

Figure 8.15 Recurrent NN [15].

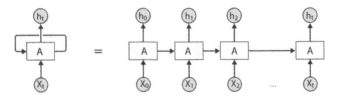

Figure 8.16 Trained model of RNN [15].

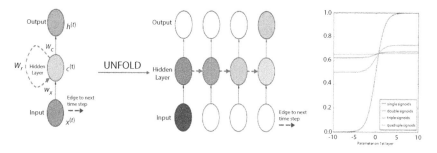

$$h^{(3)} = \sigma(w_c \cdot c^{(3)})$$

$$h^{(t)} = \sigma(w_c \cdot c^{(t)})$$

$$c^{(t)} = \sigma(w_r \cdot c^{(t-1)} + w_x \cdot x^{(t)})$$

$$= \sigma(w_c \cdot \sigma(w_x \cdot x^{(3)} + w_r \cdot c^{(2)}))$$

$$= \sigma(w_c \cdot \sigma(w_x \cdot x^{(3)} + w_r \cdot \sigma(w_x \cdot x^{(2)} + w_r \cdot c^{(1)})))$$

$$= \sigma(w_c \cdot \sigma(w_x \cdot x^{(3)} + w_r \cdot \sigma(w_x \cdot x^{(2)} + w_r \cdot \sigma(w_x \cdot x^{(1)} + w_r \cdot c^{(0)}))))$$

Figure 8.17 RNN flow [15].

Central Idea:
A memory cell (interchangeably block) comprises an explicit memory (also known as the cell state vector) and gating units that control the flow of information into and out of the memory as you can see in Figure 8.18 which represents the LSTM network.
Cell State Vector:

- Represents the memory of the LSTM
- Changes forgetting of old memory (forget gate) and addition of new memory (input gate)

Gates:

- Gate control the flow of information to/from the memory

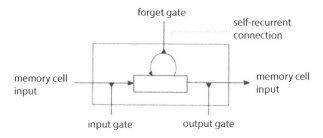

Figure 8.18 LSTM network [15].

- Gates are controlled by concatenating the output from the previous time step, the current input, and the vector of the cell state.

Forget Gate

- Determines what information should be discarded from memory.

Input Gate

- Determines what additional information is added to the cell state as a result of the current input as illustrated in Figure 8.19.

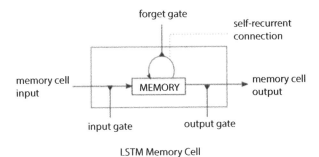

Figure 8.19 Flow chart of LSTM [15].

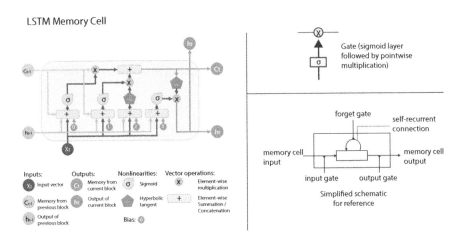

Figure 8.20 LSTM memory cell [15].

Memory Update

- The cell state vector combines the two elements (old memory via the forget gate and new memory via the input gate)

Output Gate

- Determines what to output from memory on a case-by-case basis as represented in Figure 8.20 LSTM memory cell.

References

1. Gandhi, V., R.D Engineering College, Artificial Intelligence.ppt, 2015, https://www.slideshare.net/vandanagandhi9/artificial-intelligenceppt-44690011.
2. Introduction to artificial intelligence (CPS 170), Spring 2009, Introduction to Artificial Intelligence Microsoft PowerPoint - cps170_intro (duke.edu), http://www.cs.duke.edu/courses/spring09/cps170/.
3. Harvey, C., A.I. vs. Machine learning vs. Deep learning: Subsets of artificial intelligence, 2019, https://www.datamation.com/artificial-intelligence/ai-vs-machine-learning-vs-deep-learning/.
4. Introduction to artificial intelligence (CPS 170), Spring 2009, Introduction to Machine Learning Microsoft PowerPoint - cps170_machine_learning (duke.edu), http://www.cs.duke.edu/courses/spring09/cps170/.
5. Data-flair, Training, https://data-flair.training/blogs/machine-learning-tutorial/.
6. Javatpoint, Artificial neural network, https://www.javatpoint.com/artificial-neural-network.
7. IBM Cloud Education, Convolution neural networks. What are convolutional neural networks?, IBM, 2020, https://www.ibm.com/cloud/learn/convolutional-neural-networks.
8. Mebsout, I., Convolutional neural network- part 1, 2020, https://www.ismailmebsout.com/Convolutional%20Neural%20Network%20-%20Part%201/.
9. Bhattacharya, A., Introduction to image classification DIP_401_lecture_7.pdf (iitb.ac.in), http://www.csre.iitb.ac.in/~avikb/GNR401/DIP/DIP_401_lecture_7.pdf.
10. Gotra, V., Text classification. Quarterly QA and Testing Expert Series - Vol 4/4, 2020, https://blog.qasource.com/what-is-text-classification.
11. Meaning Cloud, What is text classification, https://www.meaningcloud.com/developer/text-classification/doc/1.1/what-is-text-classification.
12. Text classification, https://monkeylearn.com/what-is-text-classification/.
13. Banerjee, P., the University of Calcutta, Image classification, https://www.learnpick.in/prime/documents/ppts/details/1638/image-classification.

14. IBM Cloud Education, Recurrent neural networks, 2020, https://www.ibm.com/cloud/learn/recurrent-neural-networks.
15. Sood, A., Long short-term memory, http://pages.cs.wisc.edu/~shavlik/cs638/lectureNotes/Long%20Short-Term%20Memory%20Networks.pdf.
16. Salma, R., (17/415128/T.K./46417), Kurniawan, A.E., (17/413537/T.K./45977), Rafi, M.F., (17/413561/T.K./46001), The benefits and the disadvantages of the artificial intelligence, https://www.coursehero.com/file/133509128/The-Benefits-and-The-Disadvantages-of-Thpdf/.
17. Nils, J., Nilsson robotics laboratory department of computer science. Introduction to machine learning, ML Book, Stanford University Stanford, CA, https://ai.stanford.edu/people/nilsson/MLBOOK.pdf.
18. A case study based on developing human artificial intelligence interaction. *JETIR*, 6, 3, March 2019.
19. Artificial intelligence and machine learning, in: *CS106E Spring 2018, Payette & Lu.*
20. Sarkar, S., CNN, Convolutional neural network (iitkgp.ac.in), 2017, http://cse.iitkgp.ac.in/~sudeshna/courses/DL17/CNN-22mar2017.pdf.
21. Introduction to machine learning: A gateway to a huge side of data science (digitalvidya.com).

9

Internet of Things

Davinder Singh

Department of Mechanical Engineering, Punjabi University, Patiala, Punjab, India

Abstract

Internet has changed drastically in the way people interact in the virtual world, in their careers or social relationships. With the increasing use of the Internet and its variety of applications, there is an increase in the number of interconnected and Internet-connected devices. Nowadays, the Internet is used in every type of organization, such as academic, research, business, personal use, etc. The Internet of Things (IoT) has attracted wide attention from researchers to address the potential of this technology in various industries recently as the current generation has become quite habitual of Internet use, and life without it seems to be impossible; everyday work and even household chores make use of the Internet. In IoT, every vital object in our daily life such as the wallet, watch, refrigerator, car, etc., will be connected with each other and with the internet. Anyone would be able to track his/her belongings from anywhere, anytime and any network. Those objects would send alerts to the users, in any event, to keep them safe. Hence, this revolution will change the way people think, live and work.

Keywords: Internet of Things (IoT), cloud computing, M2M, wireless sensor network (WSN), fog computing

9.1 Introduction

Recent years have witnessed the growing use of Internet as billions of people browse the web to access multimedia content and services, send and receive electronic mails, play games, and perform various tasks [1]. Internet has changed drastically in the way people interact in the virtual world, in their careers or social relationships [2]. This change creates a

Email: davinder5206@yahoo.co.in

Chandan Deep Singh and Harleen Kaur (eds.) *Factories of the Future: Technological Advancements in the Manufacturing Industry*, (195–228) © 2023 Scrivener Publishing LLC

global platform for machines and smart objects to communicate, dialogue, compute, and coordinate [1], which in turn builds up a strong connection among the users of smart devices worldwide. With the increasing use of the Internet and its variety of applications, there is an increase in the number of interconnected and Internet-connected devices. Nowadays, the Internet is used in every type of organization, such as academic, research, business, personal use, etc. There is rapid development and enhancements in the field of Internet which provides motivation for further research in the same or connected domains [3].

Many things in our daily lives have become wirelessly connected with miniatures and low-powered wireless devices such as radio-frequency identification (RFID) tags, near-field communication (NFC), Electronic Product Code (EPC), mobile devices and GPS, along with the development of many technologies such as sensors, actuators, embedded computing, machine-to-machine (M2M) communication and cloud computing. It was expected that there will be 7 trillion wireless devices to serve 7bn people (1,000 devices/person). These ultra-huge numbers of connected devices or things and the new techniques have introduced a paradigm, commonly referred to as the Internet of Things (IoT) [4]. IoT technology has added a new vision to this process by facilitating connections between smart objects and humans, and also between smart objects themselves, which leads to anything, anytime, anywhere, and any media communications. IoT allows objects to physically see, hear, think, and perform tasks by making them talk to each other, share information and coordinate decisions [2].

The IoT has attracted wide attention from researchers to address the potential of this technology in various industries recently as the current generation has become quite habitual of Internet use, and life without it seems to be impossible; everyday work and even household chores make use of the Internet. It seems technology is gradually becoming human-centric over the years. IoT is considered to be the future internet, which is significantly different from the Internet we use today. In IoT, every vital object in our daily life such as the wallet, watch, refrigerator, car, etc., will be connected with each other and with the internet. Anyone would be able to track his/her belongings from anywhere, anytime and any network. Those objects would send alerts to the users, in any event, to keep them safe. Hence, this revolution will change the way people think, live and work [5].

The Internet of Things (IoT) which is defined as the "digitization of machines, vehicles, and other elements of the physical world" [6] has received enormous attention from academia and various industries over the past decade. It is growing at an exponential rate as the internet is engaging people in various technologies embedded in smart devices, and the IoT is bringing fundamental changes in economic, environmental, healthcare, social and political realms in the developing world [7]. The term "Internet of Things" (IoT) came into existence when Kevin Ashton used it for the first time in 1999 to represent the globally emerging Internet-based information service architecture [8]. Weber [9] defined IoT as *"an emerging global, Internet-based information service architecture facilitating the exchange of goods in global supply chain networks … on the technical basis of the present Domain Name System; drivers are private actors"*.

IoT can also be defined over three perspectives, namely, the perspective of things or the devices that will be used as the sensing objects, the perspective of Internet or a uniform framework to which all objects are connected, and the perspective of semantics or the communication protocols over which the processing occurs [10]. IoT is considered to be a method of universal computing over devices with unique addressing schemes, having the capability to communicate and exchange data among themselves [11]. IoT will enable the objects and people using those objects to be connected in any circumstances, including any place, at any time with anyone or anything, and work on any network or path or service or communication mechanism [12]. IoT has been defined as an open network of object's capacity to share resources and data, organize, react, and respond to changes or circumstances in the surroundings, automatically [13].

IoT is an expansion of Mark Weiser's vision of ubiquitous computing (UbiComp), which aims to produce a global network that supports UbiComp and context awareness among devices. Establishing IoT is based on the proliferation of wireless sensor network (WSN), mobile computing (MobiComp), UbiComp and information technologies. Ambient intelligence is one of the key components of IoT. UbiComp and context awareness are the main requirements of ambient intelligence, in which everyday devices would be able to understand their environment, interact with each other and with humans and make the decisions [14–16].

IoT facilitates a safe and trustworthy way of exchanging information related to goods and services in a global supply chain. It acts as a pillar for ubiquitous computing that opens the door for smart environments to spot and track items, and collects information from the Internet for their proper functioning. In doing so, members of MIT Auto-ID Center developed Electronic Product Code (EPC) that serves as a universal identifier for any specific item [17, 18]. The main objective behind this development was to spread awareness about the use of Radio-Frequency Identification (RFID) globally. But, these days, the idea of "Thing" is not only restricted to RFID. It has expanded to include any real or physical object (e.g., RFID, sensor, actuator, and smart item), "spime" data object as well as any virtual or digital system, which is capable of moving in time and space. These entities can be identified uniquely through the identification details (numbers, names and/or location addresses) assigned to them. Thus, the "Thing" can be read, recognized, located, addressed and controlled effortlessly by using Internet [19].

9.2   M2M and Web of Things

Both M2M and IoT are results of the technological progress over the last decades, including not just the decreasing costs of semiconductor components, but also the spectacular uptake of the Internet Protocol (IP) and the broad adoption of the Internet. The application opportunities for such solutions are limited only by our imaginations; however, the role that M2Mand IoT will have in industry and broader society is just starting to emerge for a series of interacting and interlinked reasons [2].

The Internet has undoubtedly had a profound impact across society and industries over the past two decades. Starting off as ARPANET connecting remote computers together, the introduction of the TCP/IP protocol suite, and later the introduction of services like email and the World Wide Web (WWW), created a tremendous growth of usage and traffic. In conjunction with innovations that dramatically reduced the cost of semiconductor technologies and the subsequent extension of the Internet at a reasonable cost via mobile networks, billions of people and businesses are now connected to the Internet [17]. Quite simply, no industry and no part of society have remained untouched by this technical revolution.

At the same time that the Internet has been evolving, another technology revolution has been unfolding _ the use of sensors, electronic tags, and actuators to digitally identify, observe and control objects in the physical world. Rapidly decreasing costs of sensors and actuators have meant that where such components previously cost several Euros each, they are now a few cents. In addition, these devices, through increases in the computational capacity of the associated chipsets, are now able to communicate via fixed and mobile networks. As a result, they are able to communicate information about the physical world in near real-time across networks with high bandwidth at low relative cost.

M2M refers to those solutions that allow communication between devices of the same type and a specific application, all via wired or wireless communication networks. M2M solutions allow end-users to capture data about events from assets, such as temperature or inventory levels.

Typically, M2M is deployed to achieve productivity gains, reduce costs, and increase safety or security. M2M has been applied in many different scenarios, including the remote monitoring and control of enterprise assets, or to provide connectivity of remote machine-type devices. Remote monitoring and control has generally provided the incentive for industrial applications, whereas connectivity has been the focus in other enterprise scenarios such as connected vending machines or point-of-sales terminals for online credit card transactions. M2M solutions, however, do not generally allow for the broad sharing of data or connection of the devices in question directly to the Internet [20].

9.3   Wireless Networks

For many years flying radio-controlled model aircraft has been a popular pastime. The plane "pilot" holds the radio control unit, which sends simple data control messages to the aircraft soaring above. The control unit is configured to transmit at a particular radiofrequency. The pilot may be displaying a colored tag, showing on which frequency the radio is operating. If someone else arrives and wants to fly a plane at the same place, then he or she must set his radio to a different frequency, or there will be interference between the two, possibly leading to a spectacular plane crash. If more people come, then even greater care must be taken, to make sure that each has a unique radio frequency. Contrast this simple image of radio data communication with a more recent setting. Picture instead a busy airport lounge: hundreds of travelers are waiting for their flights. Many are on their mobile phones; others have laptops or tablets in use, maybe with wireless-linked mice, keyboards or headphones. Potentially thousands of data messages are flying through the air. All must get through reliably; none should interfere with any other [21].

Wired or *wireless*? Let's examine just why we'd want to connect without wires, and what it might offer us in tangible terms; we can use the paradigm of our own personal area network (PAN).We have a PC with its ubiquitous mouse and keyboard, a laptop, a personal digital assistant (PDA), a mobile phone with a "hands free" kit and a printer. How do we currently communicate between these devices? The answer is: with a rather unwieldy network of cables, hubs, and connectors – plugging, unplugging, and synchronizing often with the compulsory intervention of the overworked and often less-than-friendly IT department!

In the wired solution scenario that we are all accustomed to, all of the mobile devices are used in the *singular* – the interaction between them is always user initiated. We generally keep our contacts' addresses in our PCs or laptops, while their phone numbers also need to be entered into our mobile phone's directory. We are effectively forced to become database managers simply in order to maintain an up-to-date record of our contact's details. We connect to our company LAN via user-initiated password entry and connect to a printer only if we have already installed the driver or have administrator rights on our PC's—nothing is unconscious.

There are many situations where we need to provide connection between different systems or subsystems. In the domestic environment, the automated household is becoming a reality. Here different household appliances and gadgets may all be connected together. Elsewhere, there are other needs for networking. The modern motor vehicle may contain dozens of embedded systems, all engaged in very specific activity, but all interconnected. In the home or car situation, connections may be long term and stable. However, other networks or connections are transitory, for example when data is downloaded from a Smartphone to a laptop over a wireless link.

Wireless technologies continue to advance in both speed and proliferation. It seems like every spring season we see merchants get frisky and revisit the prospects of pushing wireless technologies down to their stores. Most big box retailers already use Wi-Fi® for inventory management and have dealt with upgrading portions of their outdated technology to comply with Payment Card Industry Data Security Standard (PCI DSS). Smaller retailers have toyed with implementing the technology to cut costs on store build-out or even to add additional registers in parking lots or other outdoor areas. To add to the capabilities of radio frequency (RF) communication, you could even build a point-of-sale (POS) network that functions over Bluetooth® or Zigbee®. We shudder to think about those kinds of networks, but it is possible and more likely to be used as the technologies mature. Wireless networks don't stop at Wi-Fi or Bluetooth, however.

Recent advancements in cellular data networks created an opportunity for a new class of card processing terminals that can process cards without Wi-Fi or hard-wired Internet connections. As we saw the effects of convergence in our cellular telephones over the last decade, we are beginning to see the same effects in the payment terminal market. Not only are the terminals becoming smaller and more functional, but some companies have gone completely paperless with their field technicians. Many companies are now giving both purpose built devices and tablet computers to their field technicians to prioritize work lists, give directions on where to go, provide traffic updates to keep their schedules efficient, keep track of spare part inventory while sending that information back to headquarters, and finally accepting a credit card for payment either via a sled or via a built-in reader.

Bluetooth

Bluetooth is a digital radio protocol, intended primarily for PAN applications, and operating in the 2.4 GHz radio band. It was developed by the Swedish phone company Ericsson, who took the name of a tenth century Viking king to name their communication protocol. Bluetooth provides wireless data links between such devices as mobile phones, wireless audio headsets, computer interface devices like mice and keyboards, and systems requiring the use of remote sensors. Bluetooth standards are now controlled by the Bluetooth Special Interest Group, though the IEEE has made an important contribution. It is a formidably complex protocol [22]. There is a core specification, and then over 40 different profiles, which specify different Bluetooth applications, for example for audio, printers, file transfer, cable replacement, and so on. There are three Bluetooth classes, based on output power, with Class 2 being the most common. The main characteristics are as follows:

- The approximate communication range is up to 100 m for Class 1 Bluetooth devices, unto 10 m for Class 2 devices, and 1 m for Class 3.
- Bluetooth is relatively low power; devices of Classes 1 to 3 use around 100, 2.5 and1 mW respectively.
- Data rates up to 3 Mbps can be achieved. Recent higher data rate versions are being adopted.
- Up to 8 devices can be simultaneously linked, in a piconet. A Bluetooth device can belong to more than one piconet.
- Spread-spectrum frequency hopping is applied, with the transmitter changing frequency in a pseudo-random manner 1600 times per second, i.e. a slot duration of 0.625 ms. When Bluetooth devices detect one another, they determine automatically whether they need to interact. Each device has a Media Access Control (MAC) address which communicating devices can recognize and initialize interaction if required.

Wi-Fi

Wi-Fi/802.11 is, at its basic level, a radio technology. This section will examine how voiceover Wi-Fi will interact with the current "reigning" champion of voice/radio technology: today's cellular networks. Wi-Fi and cellular are for the most part complementary technologies. They individually provide solutions to a subset of the wireless space. In particular, Wi-Fi can be used to provide coverage in areas where cellular is less effective: indoors, hospitals, airports, and urban canyons. The inclusion of a voice-over-Wi-Fi capability into a cell-phone handset can be viewed as the ultimate goal for voice over Wi-Fi. There are several reasons for this goal. An obvious one is economy of scale. Economically, a voice-over-Wi-Fi implementation (chip set/software solution) will reap immense benefits from deployment in the 500-million plus cellular-handset market. Such volumes allow for the research and development necessary to create new classes of system on a chip specifically tailored to meet conventional cellular and Wi-Fi requirements. We should expect chip sets in future that include Wi-Fi radios, MAC/Baseband integrated with cellular modems and processors. With this integration comes a reduction in cost that will surely spill over into pure voice-over-Wi-Fi space as well.

Secondly, like its parent technology VoIP, voice over Wi-Fi will piggyback on the ongoing improvements to the data networks that are being upgraded to provide enhanced data services to the basic cell phone. Examples of this include the 3G data networks that have been coming online in the past five years.

A third factor is the usefulness of Wi-Fi technology to augment areas where cellular technology is lacking. Specifically, areas where cellular coverage is problematic (in doors, airports, hospitals, and urban canyons) can be handled by overlapping Wi-Fi networks. The introduction of Wi-Fi-based mesh networks, driven by the maturation of 802.1 Is, will contribute to this trend.

For example, vendor studies have shown that a city-wide, Wi-Fi mesh network can be a more cost-effective approach to providing wireless coverage than deploying 3G (IxEV-DO). A Wi-Fi mesh network could be, for example, situated in street lamp posts (which can be rented cheaply from the city government) as opposed to a cell tower that would require space from an office building or home.

A fourth factor is bandwidth. The cellular network providers would love to have the ability to move customers off their precious spectrum wherever possible. For example, when in your broadband-enabled home, why not let the cell phone make calls over the IP-based broadband network, via your in-home Wi-Fi infrastructure? A side effect of this is that cellular providers, riding on top of broadband access, now have a means to get customer phone minutes when he is at home. By providing an integrated access point and VoIP gateway equipment that allows the customer's conventional home telephony (i.e., POTs phones) to place VoIP calls back to the cellular base network, just like with his dual-mode cell phone, the cellular providers can cut into the traditional home voice-service monopoly held by the local telephone companies [23].

This is one of the key drivers of fixed/mobile convergence, with the goal being to get customers to sign up to an all-inclusive home and mobile phone service from the cellular providers. A final factor is the cellular world trend to move to using an SIP-based call-signaling protocol known as IP Multimedia Subsystems, or IMS. We will discuss IMS further below. The use of SIP to control cell-phone call signaling as well as voice-over-Wi-Fi signaling makes it easy(relatively speaking) to architect a unified phone with seamless handoffs between the cell world and the Wi-Fi world.

9.4   Service Oriented Architecture

Architecture can be described in several different views to capture specific properties that are of relevance to model, and the views chosen in this book are the functional view, deployment view, process view, and information view. The topic of architecture is the subject of Chapter 6 of this book where more details of the definitions and purpose of architecture, as well as state of the art examples from M2M and IoT, are provided. When creating a model for the reference architecture, one needs to establish overall objectives for the architecture as well as design principles that come from understanding some of the desired major features of the resulting system solution. For instance, an overall objective might be to decouple application logic from communication mechanisms, and typical design principles might then be to design for protocol interoperability and to design for encapsulated service descriptions. These objectives and principles have to be derived from a deeper understanding of the actual problem domain, and is typically done by identifying recurring problems or type solutions, and thus by that, extracting common design patterns. The problem domain establishes the foundation for the subsequent solutions. It is common to partition the architecture work and solution work into two domains, each focusing on specific issues of relevance at the different levels of abstraction [24].

Within existing work for deriving requirements and creating architectures or reference models for IoT and M2M, three primary sources can be identified. Two of them are the larger European 7th Framework Program research projects, SENSEI (2013) and IoT-A (2013), the third being the result of a standardization activity driven by ETSI in their Technical Committee (TC) M2M (ETSI M2M TC 2013). These sources have been selected, as they represent state-of-the-art in terms of creating more complete architectures for the IoT and M2M. The approach taken in SENSEI was to develop an architecture and technology building block that enable "Real World integration in a future Internet." Key features include the definition of a real world services interface and the integration of numerous Wireless Sensor and Actuator Network deployments into a common services infrastructure on a global scale. The service infrastructure provides a set of services that are common to a vast range of application services and is separated from any underlying communication network for which the only assumption made was that it should be based on Internet Protocol (IP) [25].

The architecture relies on the separation of resources providing sensing and actuation from the actual devices, a set of contextual and real world entity-centric services, and the users of the services. SENSEI further relies on an open-ended constellation of providers and users, and also provides a reference model for different business roles. A number of design principles and guidelines are identified, and so is a set of requirements. Finally, the architecture itself contains a set of key functional capabilities. The telecommunications industry, meanwhile, has focused on defining common service core for supporting various M2M applications, and that is agnostic to underlying networks in ETSI TC M2M. The approach taken has been to analyze a set of M2M use cases, derive a set of M2M service requirements, and then to specify architecture as well as a set of supporting system interfaces. Similar to SENSEI, there was a clear approach towards a horizontal system with separation of devices, gateways, communications networks, and the creation of a common service core and a set of applications, all separated by defined reference points.

Finally, the approach taken in IoT-A differs from the two approaches above in the sense that instead of defining a single architecture, a reference architecture is created, captured in what the IoT-A refers to as the Architectural Reference Model (ARM). The vision of IoT-A is, via the ARM, to establish a means to achieve a high degree of interoperability between different IoT solutions at the different system levels of communication, service, and information. IoT-A provides a set of different architectural views, establishes a proposed terminology and a set of Unified Requirements (IoT-A UNI 2013). Furthermore, IoT-A proposes a methodology for how to arrive at a concrete architecture based on use cases and requirements [24].

9.5   Complexity of Networks

Generally speaking, within the wider context of innovation, sensor networks will play an increasingly important role: environmental control of the passenger cabin and connecting personal electronic equipment, helping pilots with decision-making, identifying the cargo and any detachable equipment (life preservers) through improved RFID tags, safe and optimal integration of the aircraft environment on the ground and in-flight.

Sensor networks will also be used for monitoring the aging of the structure of the aircraft as well as its components. Essentially, an aircraft often finds itself in a harsh environment (variations in temperature and pressure, humidity, vibrations, dust, collisions, etc.), while for some of its components there are added specific constraints (acoustic charge and temperature in the vicinity of the engine and the brakes, mechanical constraints, corrosive liquids, etc.) It is therefore essential to carry out regular inspections and maintenance. Traditionally, these are carried out periodically, looking to anticipate predictable faults or decreased efficiency. Health Monitoring by sensor networks will make it possible to only carry out repairs when necessary, thus saving time (especially for areas which are difficult to access), weight (by decreasing mechanical margins) and therefore fuel [26].

9.6   Wireless Sensor Networks

Wireless Sensor Network (WSN) is perception, and the main tool is sensor. The use of sensors can effectively sense the external environment, and then with the help of wireless network for information transmission, to meet user needs. The security of wireless sensor network must be paid attention to, because it is supported by high technical content and complex structure, once there is a problem, the consequences are unimaginable. The development of the Internet of things is inseparable from wireless sensor networks, and the ability of people to communicate quickly is also closely related to it [23]. WSN can be used to collect all kinds of data and information and deal with all kinds of complex environments, the application of WSN covers all aspects. In the military field, it can be used to detect the deployment of troops in enemy terrain and other situations, and it has the function of detecting biological and chemical pollution and nuclear radiation. In terms of environmental monitoring and protection, it can be used for data collection in the field environment, tracking animal footprints, analyzing pollution situation, and predicting the explosion of forest fire and debris flow. In the field of industry and agriculture, the monitoring of crop growth and product flow intelligent production. In the field of medical care and health, it can realize the function of obtaining physiological data of patients, analyzing their conditions, and timely receiving medical treatment. In addition, wireless sensor network plays a crucial role in the field of smart home, smart transportation, smart city, cultural relic protection and business, etc. Wireless sensor network is changing people's life [27].

The information security of WSN is concerned in many contexts, such as military medical disaster prevention and other fields. Wireless sensor networks use open wireless communication channel technology to transmit data, but without security protection means, data is very vulnerable to internal and external attacks [28]. However, the amount of computation based on the cryptographic defense method is not suitable for WSN network. Therefore, it is also a big problem to choose an appropriate encryption method to ensure the security of WSN information [29].

A wireless sensor network draws together systems (unfairly termed *sensors*) which comprise the nodes of this network. There can be a large number of these nodes or a relatively small number. Each node must be able to transmit information, gathered by its sensor(s), to a collection point.

The transmission can be direct or can move around and in this case each node (or a few) plays the role of a relay and eventually a router. In the case of a radio link, transmission *via* relay makes it possible to reduce the power radiated by the antenna, given that the power received by the antenna is a function of the inverse square of the distance between nodes3. Although this is tempting, the use of a relay is nevertheless a complicated method in terms of synchronizing the sending of frames, the management of collisions and the routing of messages.

Finally, some nodes can be used for aggregating data, in order to reduce transmission flows, and to avoid overcharging the relay nodes located at the vicinity of the collection point. Nodes will therefore not necessarily be identical.

The network can be deployed manually, where each node occupies a well-defined position, or by chance (natural dispersion). The topology of the network can be fixed (deployment in a building, a boat, an aircraft or a factory) or can evolve over time (deployment over several mobile objects, in a liquid environment, on cooperative drones). In situations where the nodes are considered inaccessible once they have been deployed (dispersed into nature, lost in a concrete structure, placed in an inaccessible area – aside from heavy disassembly – of an aircraft), it will be necessary to consider tolerance of errors and self-organization.

Security of WSN

Security of wireless sensor networks (WSN) has become a hotspot in the research of WSN technology. The characteristics of wireless sensor networks make them vulnerable to multiple attacks, which seriously affects the confidentiality, integrity and availability of data. Especially when the routing of the network is attacked, the data collected by the sensor node cannot be transmitted to the destination sink node timely and accurately. Just like the traditional computer communication network, the security threat of wireless sensor network mainly comes from various attacks. Radio characteristics of wireless channel and ad-hoc network features are making it easier for the sensor network attacker's passive attacks and active attacks, sensor networks will be monitored, tampered with, forge and block attacks, at the same time, wireless sensor network as a kind of energy depletion, energy constrained sensor nodes, the network is vulnerable to denial of service attacks [20].

Combined with the characteristics of wireless sensor network, the security purpose of wireless sensor network is summarized based on the analysis of security threats at all levels of the network: confidentiality, to prevent data eavesdropping and stealing by illegal users, key management to provide a secure key update and management mechanism; Integrity of data to prevent information from being tampered with illegally; Freshness of data to prevent malicious nodes from sending the same information to consume network resources; Auditability, can audit the whole access process of the system; Non-repudiation, the sensor nodes participating in the network communication process can't deny its behavior; Access authentication, to verify the sensor nodes in the network, to verify the legitimacy of its identity; The physical device is safe to prevent sensor nodes from being stolen, destroyed, and safe to use [24].

9.7   Cloud Computing

In the past few years, clouds have emerged as effective computing platforms to face the challenge of extracting knowledge from big data repositories, as well as to provide effective and efficient data analysis environments to both researchers and companies. From a client perspective, the cloud is an abstraction for remote, infinitely scalable provisioning of computation and storage resources. From an implementation point of view, cloud systems are based on large sets of computing resources, located somewhere "in the cloud", which are allocated to applications on demand [30].

With the rapid development of processing and storage technologies and the success of the Internet, computing resources have become cheaper, more powerful and more ubiquitously available than ever before. This technological trend has enabled the realization of a new computing model called cloud computing, in which resources (e.g., CPU and storage) are provided as general utilities that can be leased and released by users through the Internet in an on-demand fashion. In a cloud computing environment, the traditional role of service provider is divided into two: *the infrastructure providers* who manage cloud platforms and lease resources according to a usage-based pricing model, and *service providers* who rent resources from one or many infrastructure providers to serve the end users. The emergence of cloud computing has made a tremendous impact on the Information Technology (IT) industry over the past few years, where large companies such as Google, Amazon and Microsoft strive to provide more powerful, reliable and cost-efficient cloud platforms, and business enterprises seek to reshape their business models to gain benefit from this new paradigm.

The definition of the National Institute of Standards and Technology (NIST)for cloud computing has been largely adopted by cloud computing communities. According to NIST, cloud computing "is a model for enabling convenient, on-demand network access to a shared pool of configurable computing resources (e.g., networks, servers, storage, applications, and services) that can be rapidly provisioned and released with minimal management effort or service provider interaction" [31]. Cloud computing is the name given to a collection of technologies such as distributed systems, grid computing, service-oriented architecture (SOA), and virtualization [32]. The main idea behind cloud computing is not a new one. John McCarthy in the 1960s already envisioned that computing facilities will be provided to the general public like a utility. The term *cloud* has also been used in various contexts such as describing large ATM networks in the 1990s [33].

However, it was after Google's CEO Eric Schmidt used the word to describe the business model of providing services across the Internet in 2006 that the term really started to gain popularity. Since then, the term cloud computing has been used mainly as a marketing term in a variety of contexts to represent many different ideas.

The main reason for the existence of different perceptions of cloud computing is that cloud computing, unlike other technical terms, is not a new technology, but rather a new operations model that brings together a set of existing technologies to run business in a different way. Indeed, most of the technologies used by cloud computing, such as virtualization and utility-based pricing, are not new. Instead, cloud computing leverages these existing technologies to meet the technological and economic requirements of today's demand for information technology. Figure 9.1 shows the main enabling technologies of cloud computing [34]. Table 9.1 identifies the essential characteristics of cloud computing [31], provides an explanation for each, and related concepts.

Distributed computing or Grid computing refers to connecting dispersed computing resources over the network in order to harness the collective power of the systems. The underlying distributed nature of the system is transparent to the end user to whom the system appears as a single supercomputer. The development of grid computing was originally driven by scientific applications which are usually computation-intensive. Cloud computing is similar to grid computing in that it also employs distributed resources to achieve application-level objectives. However, cloud computing takes one step further by leveraging virtualization technologies at multiple levels (hardware and application platform) to realize resource sharing and dynamic resource provisioning.

Data center refers to a facility with large number of computers and storage devices interconnected with high-speed network lines [35]. Data centers construct the physical infrastructure on which cloud computing services operate. These centers have specific security, power management, and cooling requirements. The hardware systems in a cloud data center are shared among many users. This is realized by virtualizing, a basic enabling technology of cloud computing, which refers to running several software-based systems (VM) in a single physical machine. Virtualization

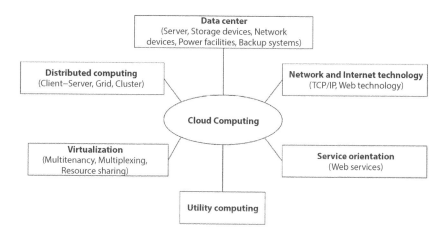

Figure 9.1 Main enabling technologies of cloud computing [34].

Table 9.1 The essential characteristics of cloud computing [31] and its related concepts.

S. no.	Characteristics related	Explanation	Concepts
1	Broad network access	Cloud services are enabled over network using Internet technology and are accessed from workstations or mobile devices, usually via a Web browser.	Distributed computing, Internet technology, datacenter
2	Measured services	Resource utilization, for example, the amount of memory or network bandwidth, is measured in detail. Users only pay for the resources they use without any upfront payment and are able to monitor and control the utilization of resources (e.g. CPU, memory, disk space).	Utility-based computing, pay-as-you-go model, no upfront costs, monitoring
3	On-demand self-service	Users can utilize services they need in an on-demand basis without any human intervention from CSP. Required functionalities are provided as services via easy to use Web pages or Application Programming Interfaces (APIs).	Service orientation, API
4	Rapid elasticity	The capacity of provisioned services and systems can be scaled up and down on demand.	Scalability, reliability, virtually unlimited resources
5	Resource Pooling	Virtualization technology allows the provision of several virtual machines (VMs) in the same physical machine. Resources, both physical and virtual, can be dynamically shared between cloud users independent of their geographic location.	Virtualization, resource sharing, multitenancy, multiplexing, energy consumption

abstracts the underlying infrastructure from the users to whom the VMs look like real physical systems. Several VMs of multiple users can share a single hardware [36], which is called multitenancy [37].

VM users can access the required functionalities provided by CSP as services like other utilities. In such on-demand and utility-based computing, the users pay for the resources they actually use. *Virtualization* is a technology that abstracts away the details of physical hardware and provides virtualized resources for high-level applications. A virtualized server is commonly called a virtual machine (VM). Virtualization forms the foundation of cloud computing, as it provides the capability of pooling computing resources from clusters of servers and dynamically assigning or reassigning virtual resources to applications on-demand. *Utility Computing* represents the model of providing resources on-demand and charging customers based on usage rather than a flat rate. Cloud computing can be perceived as a realization of utility computing. It adopts a utility-based pricing scheme entirely for economic reasons. With on-demand resource provisioning and utility-based pricing, service providers can truly maximize resource utilization and minimize their operating costs.

In summary, cloud computing leverages virtualization technology to achieve the goal of providing computing resources as a utility. It shares certain aspects with grid computing and autonomic computing but differs from them in other aspects. Therefore, it offers unique benefits and imposes distinctive challenges to meet its requirements.

9.8 Cloud Simulators

Nowadays, cloud computing is an emerging technology due to virtualization and providing low price services on pay-as per-use basis. One of the main challenges in this regard is how to evaluate different models of the cloud resources usage and the ability of cloud systems based on QoS constraints. Experimentation in a real environment is very difficult and expensive. Therefore, many works pay much attention on designing cloud simulation frameworks that only cover a subset of the different components. Consequently, choosing the right tools to use needs a deep comparative analysis of available simulators based on different features [38]. Also, the Cloud to Fog continuum is a very dense and complex scenario. At the core level (Cloud) resources are vast, whilst they become scarce at the Edge (Fog). This complexity leads to the need of simulation tools in order to evaluate the performance of novel mechanisms that hardly can be tested in real scenarios. Thus, simulation represents a solution for early stage evaluation before moving to real-world (and more expensive and complex) test beds. However, selecting the appropriate simulation tool can be complex in itself [25].

The importance of cloud simulators

Each cloud system should have some characteristics such as the awareness of geographic location, high access time, rapid elasticity, management of heterogeneity data, and application scalability to achieve its primary goals [39]. These features inherently lead to the development of complex application modules and resource management schema. New strategies that are designed for the cloud to take advantage of the characteristics of the cloud must be investigated in various aspects like low latency [39, 40]. Most of the cloud researchers have been interested to study the cloud challenges like power saving, cost modeling, load balancing, and security concepts. Recently, it becomes more and more significant to discuss around some areas of cloud computing such as:

- How do cloud computing and cloud-based applications perform?
- Are services of cloud computing safe and privacy protected?
- Which cloud services are more energy-efficient and affordable?

Different systems collect large amounts of data independently of the human business processes and then deploy this information in IT and business purposes. When there are several parameters relevant to technical performance, then proposing an efficient cloud solution is becoming more and more challenging. Therefore, developers require an appropriate way to model their strategies and extend data management policies. Testing in a real environment is expensive, time consuming, and unrepeatable especially for performance (e.g., throughput, cost benefits) and security issue analysis of cloud. In other words, quantifying the performance of scheduling strategies in a real cloud under transient conditions will lead to many problems for the following reasons:

- Cloud consists of different requests, access patterns, and resources (i.e., hardware, software, and network).
- Users show dynamic behavior based on QoS requirements.
- Applications may have varying workloads and characteristics.

- Cloud-based environment control is not in the hands of developers and so it is not feasible to repeat benchmarking experiments under the same conditions.

In terms of cost factors, using of real resources for the evaluation of new strategies from the beginning of the solution development process is not always feasible. On the other hand, using of benchmark solution (i.e., considering a set of servers) does not ensure a suitable view of scalability issues. While scalability plays an important role in cloud scenarios and evaluation by analytical methods is impossible due to the increasing complexity.

As a result, the simulator can play an important role in reducing cost, efficiency, infrastructure complexity, and security risks before a solution can be deployed on real infrastructure. By focusing on issues related to the quality of a particular component under various scenarios, cloud simulators enable performance analysts to monitor the behaviors of the system. In summary, cloud simulators have the following general advantages compared to cloud service [41]:

- Cloud simulators do not need installation and the costs of maintenance (i.e., no investment). In addition, risk assessment in simulation tools during the early stages does not involve the capital cost while it is implemented. Thus, a developer can recognize a risk that is related to the design or any parameter.
- In cloud simulators, user can easily change inputs, scenarios, and load. Therefore, the results as output can be analyzed in different conditions.
- Simulator learning is easy for developers. If they know the programming language well, the evolution with simulators will not be a problem.

Furthermore, potential users who benefit from cloud simulators include:

- Cloud providers and architecture: They are the main users and can design and evaluate their strategies with a simulator.
- Cloud clients: Simulators can be useful for large companies to compare various providers of private and public clouds, evaluate cloud deployment solutions, and study customer-related workloads.

- Scientific community: Simulators are primary tools for researchers to analyze their proposed methods before testing on a real setup.
- External players: The simulators can be helpful for everyone who is worried about cloud applications. For example, government employees should be concerned about energy efficiency or carbon printing from a cloud. Therefore, their researchers can study energy performance configuration with a suitable cloud simulator.

Without cloud simulators, researchers and cloud providers only have to use theoretical and inaccurate assessment or trial and error that lead to a waste of resources. In the literature, there are traditional simulators for distribution systems that do not have the ability to model the cloud environment. In the last few years, different cloud simulators are designed for evaluating resource utilization [42].

9.9   Fog Computing

With the increasing advancement in the applications of the Internet of Things (IoT), the integrated Cloud Computing (CC) faces numerous threats such as performance, security, latency, and network breakdown. With the discovery of Fog Computing these issues are addressed by taking CC nearer to the Internet of Things (IoT). The key functionality of the fog is to provide the data generated by the IoT devices near the edge. Processing of the data and data storage is done locally at the fog node rather than moving the information to the cloud server. In comparison with the cloud, Fog Computing delivers services with high quality and quick response time. Hence, Fog Computing might be the optimal option to allow the Internet of Things to deliver an efficient and highly secured service to numerous IoT clients. It allows the administration of the services and resource provisioning outside CC, nearer to devices, at the network edge, or ultimately at places specified by Service Level Agreements (SLA's). Fog Computing is not a replacement to CC, but a prevailing component. It allows the processing of the information at the edge though still delivering the option to connect with the data center of the cloud.

We define Fog computing as a distributed computing paradigm that fundamentally extends the services provided by the cloud to the edge of the network (as shown in Figure 9.2). It facilitates management and programming of compute, networking, and storage services between data

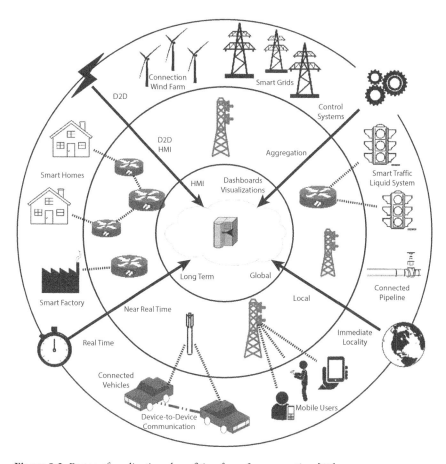

Figure 9.2 Range of applications benefiting from fog computing [47].

centers and end devices. Fog computing essentially involves components of an application running both in the cloud as well as in edge devices between sensors and the cloud that is, in smart gateways, routers, or dedicated fog devices. Fog computing supports mobility, computing resources, communication protocols, interface heterogeneity, cloud integration, and distributed data analytics to address requirements of applications that need low latency with a wide and dense geographical distribution.

Fog computing is defined as "an extremely virtualized environment that delivers networking, storage, and compute resources between outdated CC information centers, usually, but not entirely situated at the network edge" [43]. A fog structure contains various edge nodes with few processing competences, which are frequently called fog nodes. These nodes of fog have less processing facilities and storage. In fog network, sometimes edge and many servers are called cloudlets [44, 45] which take part in the shared computing surroundings, not outside the network edge. By using these devices of fog, the clients might obtain a real-time response for sensitive latency applications. Even though the phrase was initially devised by Cisco [43], various researchers and industries defined fog computing from many different perspectives. Abroad spectrum of Fog computing is specified as "geographically shared computing framework with a pool of requirements that contains different universally linked heterogeneous computing devices at the network edge and not entirely flawlessly supported by services of cloud to collectively offer transmission, storage and elastic computation in remote surroundings to an enormous scale of users in closeness" [46].

Advantages associated with fog computing includes the following:

- *Reduction of network traffic:* Cisco estimates that there are currently 25 billion connected devices worldwide, a number that could jump to 50 billion by 2020. The billions of mobile devices such as smart phones and tablets already being used to generate, receive, and send data make a case for putting the computing capabilities closer to where devices are located, rather than having all data sent over networks to central data centers. Depending on the configured frequency, sensors may collect data every few seconds. Therefore, it is neither efficient nor sensible to send all of this raw data to the cloud. Hence, fog computing benefits here by providing a platform for filter and analysis of the data generated by these devices close to the edge, and for generation of local data views. This drastically reduces the traffic being sent to the cloud.
- *Suitable for IoT tasks and queries:* With the increasing number of smart devices, most of the requests pertain to the surroundings of the device. Hence, such requests can be served without the help of the global information present at the cloud. For example, the aforementioned sports-tracker application Endomondo allows a user to locate people playing a similar sport nearby.
- Because of the local nature of the typical requests made by this application, it makes sense that the requests are processed in fog rather than cloud infrastructure. Another example can be a smart-connected vehicle which needs to capture events only about a hundred meters from it. Fog computing makes the communication distance closer to the

physical distance by bringing the processing closer to the edge of the network.

- *Low-latency requirement:* Mission-critical applications require real-time data processing. Some of the best examples of such applications are cloud robotics, control of fly-by-wire aircraft, or anti-lock brakes on a vehicle. For a robot, motion control depends on the data collected by the sensors and the feedback of the control system. Having the control system running on the cloud may make the sense-process-actuate loop slow or unavailable as a result of communication failures. This is where fog computing helps, by performing the processing required for the control system very close to the robots—thus making real-time response possible.
- *Scalability:* Even with virtually infinite resources, the cloud may become the bottleneck if all the raw data generated by end devices is continually sent to it. Since fog computing aims at processing incoming data closer to the data source itself, it reduces the burden of that processing on the cloud, thus addressing the scalability issues arising out of the increasing number of endpoints.

As demonstrated in Figure 9.2, there is a variety of applications benefiting from the Fog-computing paradigm [47].

Initially, the fog was introduced to serve the IoT applications [48], many applications based on Wireless Sensor Network (WSN) started to support fog computing. Almost all applications that have latency as an issue started to take advantage of the fog environment. These included any type of utility services that could integrate with fog to provide better service and minimize cost. In an application, in which the system uses Augmented Reality can adopt fog infrastructure as it will change the current world in the future. The requirements of processing in real-time using augmented reality can be catered by fog environment which can cause prolonged improvement in many services of augmented reality.

9.10   Applications of IoT

The possible applications offered by the IoT are numerous and diverse in all areas of everyday life. IoT new applications are likely to improve the quality of our daily lives in many areas and environments: at home, workplace, hospital, gym, while traveling, just to name a few. With IoT, these

Table 9.2 IoT applications [2].

S. no.	Domain	Applications
1	Community domain	Smart home/smart building Smart grids and smart metering Smart cities Security and surveillance Healthcare Social IoT (SIOT)
2	Environment domain	Smart agriculture and smart water Animals and breeding Environmental monitoring Recycling
3	Transportation domain	Intelligent transportation systems (ITS) or transportation cyber-physical systems (T-CPS) Robot taxi
4	Industry and manufacturing domain	Industrial automation Oil and gas industry Supply chains/logistics Pharmaceutical industry Insurance industry
5	Entertainment industry	Enhanced game room Media Music cognition

environments would be equipped with intelligent objects that are able to communicate with each other and exchange the captured information; therefore, a very wide range of applications can be deployed. IoT application areas are limited only by imagination [15, 49]. In this chapter, IoT applications are categorized according to their domains into five groups: community, transportation, environmental, industrial and entertainment domains. Table 9.2 shows the applications in each category, which will be explained in the next subsections.

Community Domain

IoT smart home services enhance personal lifestyle by facilitating remote monitoring of home appliances and systems such as energy consumption meters, air conditioners, heating systems, etc., and performing operations remotely and automatically, for example, closing the windows automatically and lowering the blinds of upstairs windows based on the weather forecast [1]. IoT smart home can help in reducing and optimizing resource consumption like electricity and water, detecting emergencies, providing home safety and security and finding things at home easily. This would help in reducing operational expenditure and carbon footprint, which would have the impact of the global greenhouse gas emission [50].

Environment Domain

IoT devices can be used in monitoring soil moisture and stem diameter in vineyards to improve and strengthen agriculture work. IoT can maximize production and improve the quality of fruits and vegetables by controlling and maintaining the amount of vitamin in agricultural products and controlling microclimatic conditions. IoT also can be used in studying weather conditions in the fields to forecast ice formation, drought, snow or wind changes, and controlling humidity and temperature levels to prevent fungus or other microbial contaminants. Smart monitoring of soil can help in making the decisions about agriculture to increase the production of food grains and prevent loss of crops. As drought threatens the environment, water conservation is a major concern, which can take benefits of smart technology. IoT can be used in studying water adequacy in rivers and sea for agricultural and drinkable use, detecting liquid presence outside tanks and pressure variations along pipes and monitoring changes in water levels in rivers, reservoirs and dams.

Transportation Domain

IoT can also be used in traffic monitoring as an important portion of smart city infrastructure. To make the transportation system reliable and smart, normal traffic and highway traffic require sufficient information about the available support and logistics. Any kind of road congestion will eventually lead to fuel and economic loss; therefore, traffic foresight will help in improving the entire system (Singh *et al.*, 2014). ITS utilizes four main components: vehicle subsystem (consists of GPS, RFID reader, OBU and communication), station subsystem (road-side equipment), ITS monitoring center and security subsystem [50]. The deployment of ITS will make the transportation of people and goods more efficient. For example, by using ITS, the containers can self-weigh and scan themselves; thus, packing containers would become more efficient. The use of IoT technologies to manage passenger luggage at airports will allow automatic tracking and sorting, increase reading rates for each bag and increase security.

Industry and Manufacturing Domain

With IoT, the industrial automation will be able to control and monitor operations, functionalities and productivity rate of production machines through the internet. The products are produced rapidly and more accurately by a group of machines based on four elements: transportation, processing, sensing and communication. IoT also raises productivity by analyzing production data and timing and production problem causes. For example, if any machine faces a sudden problem, IoT system will immediately send a maintenance request to the maintenance department to handle and fix the issue [50].

Entertainment Industry

Game rooms and players will be equipped with various devices and sensors for tracking the information of movement, acceleration, humidity, temperature, noise, sound, visual information, location sensitivity, blood pressure and heart rate. By using this information, the game room can measure the excitement and energy level of players to control the game activity depending on the players' situations [49]. IoT technology could be queried for news gathering. The information would be collected from multimedia devices available in a specific location, and the multimedia footage about a certain event would be sent to the news station. Another example is attaching NFC tags to the posters to provide more information by connecting the reader to a URI address that contains more detailed information about the poster [51].

9.11 Research Gaps and Challenges in IoT

There are many challenges that need to be addressed for better achievement of IoT vision [50]. In order to ensure the adoption and dissemination of IoT, these gaps must be overcome appropriately. These gaps are summarized in the next subsections.

Architecture

TCP/IP protocol stack has been used in today's network hosts communication for a longtime. Nevertheless, the TCP/IP protocol stack is not suitable for IoT. IoT will connect billions of objects, in which everything and everyone would exchange information; thus, much more traffic will exponentially be generated and more data storage will be needed. Therefore, the new proposed architecture for IoT has to address many issues such as scalability, interoperability, reliability, QoS, etc. [52]. Since IoT includes various smart devices and sensors with a wide range of technologies, single reference architecture cannot be used as a blueprint for all applications [53]. Moreover, the choice of IoT architecture itself is a big challenge that paves the way for the development of a new architecture and the modification of the current architecture [5].

Standardization

Many manufacturers provide devices that use their proprietary technologies and services and to which others may not have access. Therefore, the standardization on IoT is very vital to provide better interoperability for all sensor devices and objects [52]. The aim of IoT standardization is to reduce the access barriers to new service providers and new users. Furthermore, standardization allows products or services to compete better at a higher level. Nevertheless, the fast growth of IoT makes the standardization difficult. Open standards for IoT such as identification standards, communication standards, security standards, etc., are some key factors in the deployment of IoT technologies [54].

Data management Challenges

IoT sensors and devices will generate a heterogeneous and huge amount of data. However, the current data center architecture is not prepared to deal with this huge volume of data appropriately. Few organizations will be able to invest in large data storage, enough for storing their IoT collected data. Approximately, more than 2.5 trillion bytes of new data will be recorded every day; therefore, data analysis will play a major role. McKinsey Global Institute estimated that the USA needs between 140,000 and 190,000 additional workers with analytical skills and 1.5m managers and analysts to make business decisions based on big data analysis. Furthermore, there will be a need for a model to integrate data semantic in order to create meaning and knowledge from this big data. AI algorithms and machine learning methods, based on genetic algorithms, evolutionary algorithms, neural networks and other AI techniques, should be applied to extract the knowledge from the redundant data and perform automated decision making [5, 55].

Cloud Computing

IoT and cloud computing integration would create a smart environment that combines multi-stakeholder services and large-scale support for a massive number of users in a decentralized and reliable way. Nonetheless, this integration needs to cope with constraints like data sources or access devices with unreliable connectivity and limited power. The cloud application platforms need to be enhanced to support the rapid creation of applications. By providing seamless execution of applications, domain-specific programming tools and utilizing the capabilities of multiple dynamic and heterogeneous resources, the cloud platforms would be able to meet the requirements of QoS of the different users [56].

Context Awareness
Understanding sensor data is one of the main challenges that would face the IoT ecosystem. The research in this area has been strongly invested and funded by the governments, interested groups, companies and research institutes like CERP-IoT. CERP-IoT, funded by the EU, had set a time frame for the development and research on IoT context awareness during 2015–2020 [57].

Quality of Service
QoS is easy to be provided in WSNs because of the ability to manage shared wireless media and the limitation of the allocated resource. Research on WSN QoS may be applied to IoT as a short-term solution. Moreover, the research on M2M communication QoS cannot be applied properly to IoT because M2M communication paradigm is different from IoT. IoT objects are various, each of which has its own behavior and characteristics. Another major research area related to QoS is cloud computing QoS, which will lead to the development of a better approach for resource allocation and management and an optimal and controlled way for serving different network traffics. Furthermore, the QoS in IoT cyber security is still an unexplored field of research [58].

Security and Privacy
IoT improves the productivity of companies and enhances the quality of people's lives. However, since IoT devices are typically wirelessly connected and may be located in public places, IoT will generate a potential surface of attacks for hackers and other cybercriminals. In addition, IoT devices have serious vulnerabilities due to insecure web interfaces, lack of transport encryption, insufficient software protection and inadequate authorization. The lack of security and privacy solutions will cause resistance to IoT adoption by companies and individuals. Security challenges can be solved by training developers to integrate security solutions, such as intrusion prevention and firewall systems, into their products and encouraging users to use the embedded IoT security features in their devices [15, 54, 55].

Managing Heterogeneity
IoT devices have different operating conditions, functionalities and resolutions. They are published by different people, authorities or entities. Therefore, seamless integration of this heterogeneity is a major challenge.

Heterogeneity should be supported at both the architectural and protocol levels [51].

Greening of IoT

IoT ecosystem will cause a significant increase in the network energy consumption due to the increase in data rates, the rapid growth of internet-connected edge devices and the rise in the number of internet-enabled services. Thus, adopting green technologies is important to make IoT devices energy efficient [52].

Scalability

The deployment of a sheer number of miniature devices like sensors and actuators and the huge volume of data produced by them pose a scalability challenge in IoT. Scalability issues arise at different levels, including naming and addressing, data communication and networking, information and knowledge management, and service provisioning and management [1].

Legal/Accountability

IoT will not be dedicated to a single group. In fact, different stakeholders will be involved. The management of IoT global paradigm will pose challenges and accountability. There is a need for a shared administration structure of IoT, which includes all the relevant stakeholders and establishes global accountability and enforcement by issuing punishments [15].

5G–IoT

5G offers advantages that can meet future IoT requirements. However, it opens up a new set of research challenges on some issues, for example, introducing reliable communication between devices that have various communication technologies. The interoperability between IoT and 5G technology has to consider cyber security issues such as data privacy, information transmission management and security protocols and mechanisms [54, 58]. In addition, the challenges in 5G-IoT architecture, NFV, D2D communications, etc., have to be addressed.

9.12   Concluding Remarks

The widespread use of miniature devices, which have the ability to communicate and actuate, is the reason that brings IoT vision. New capabilities are made possible as the sensors and actuators function and run smoothly in the background and access rich sources of massive information. The creativity of users in designing new applications will control the evolution of the next-generation mobile phone system. IoT paradigm is ideal for influencing that field by providing new data and computational resources required to create revolutionary applications. IoT devices will enable ambient intelligence and they will be pervasive, ubiquitous and context aware. IoT will improve human being's lives through automation and augmentation. IoT capabilities can save time and money of organizations and people by helping them in making the proper decisions fast. The ground on which the IoT ecosystem was built is not new. In reality, IoT utilizes existing technologies such as WSNs and RFID along with standards and protocols for M2M communication. IoT has the potential ability to reshape our world by combining existing technologies in a newer way. The question remains whether IoT will be a permanent technology, fail to be achieved, or be the starting point of another model. Time alone will eventually answer these questions.

References

1. Miorandi, D., Sicari, S., Pellegrini, F.D., Chlamtac, I., Internet of things: Vision, applications and research challenges. *Ad Hoc Networks*, 10, 1497–1516, 2012.
2. Shammar, E.A. and Zahary, A.T., The Internet of Things (IoT): A survey of techniques, operating systems, and trends. *Internet Things (IoT)*, 38, 1, 5–66, 2019.
3. Xu, X., Huang, S., Chen, Y., Browny, K., Halilovicy, I., Lu, W., TSAaaS: Time series analytics as a service on IoT. *2014 IEEE International Conference on Web Services*, pp. 249–256, 2014.
4. Razzaque, M.A., Milojevic-Jevric, M., Palade, A., Clarke, S., Middleware for Internet of Things: A survey. *IEEE Internet Things J.*, 3, 1, 70–95, 2016.
5. Singh, D., Tripathi, G., Jara, A.J., A survey of Internet-of-Things: Future vision, architecture, challenges and services. *2014 IEEE World Forum on Internet of Things*, pp. 287–292, 2014.
6. Manyika, J., Chui, M., Bisson, P., Woetzel, J., Dobbs, R., Bughin, J., Aharon, D., *The Internet of Things: Mapping the value beyond the hype*, McKinsey Global Institute Report, San Francisco, CA, 2015.
7. Kshetri, N., The economics of the Internet of Things in the global South. *Third World Q.*, 38, 2, 311–339, 2017.
8. Ashton, K., That 'Internet of Things' thing. *RFID J.*, 22, 7, 97–114, 2009.
9. Weber, R.H., Internet of Things – Need for a new legal environment? *Comput. Law Secur. Rev.*, 25, 6, 522–527, 2009.
10. Yang, D.L., Liu, F., Liang, Y.D., A survey of the internet of things, in: *International Conference on E-business Intelligence*, Atlantis Press, 2010.
11. Agrawal, S. and Vieira, D., A survey on Internet of Things. *Abaks*, 1, 2, 78–95, 2013.

12. Kumar, J.S. and Patel, D.R., A survey on Internet of Things: Security and privacy issues. *Int. J. Comput. Appl.*, 90, 11, 20–26, 2014.

13. Madakam, S., Ramaswamy, R., Tripathi, S., Internet of Things (IoT): A literature review. *J. Comput. Commun.*, 03, 05, 164–173, 2015.

14. Porkodi, R. and Bhuvaneswari, V., The Internet of Things (IOT) applications and communication enabling technology standards: An overview. *2014 International Conference on Intelligent Computing Applications*, IEEE, pp. 324–329, 2014.

15. Whitmore, A., Agarwal, A., Da Xu, L., The Internet of Things – A survey of topics and trends. *Inf. Syst. Front.*, 17, 2, 261–274, 2015.

16. Zhang, D., Yang, L.T., Huang, H., Searching in Internet of Things: Vision and challenges. *2011 IEEE Ninth International Symposium on Parallel and Distributed Processing with Applications*, pp. 201–206, 2011, available at: https://doi.org/10.1109/ISPA.2011.53.

17. Gama, K., Touseaui, L., Donsez, D., Combining heterogeneous service technologies for building an Internet of Things middleware. *Comput. Commun.*, 35, 4, 405–417, 2012.

18. EPC Global Inc., GS1 EPC Tag Data Standard 1.6, EPC Global, 1, 1–218, 2011, <http://www.gs1.org/gsmp/kc/epcglobal/tds/tds16-RatifiedStd-20110922.pdf>.

19. Borgia, E., The Internet of Things vision: Key features, applications and open issues. *Comput. Commun.*, 54, 1, 1–31, 2014.

20. Lee, H.M., Optimal cost design of water distribution networks using a decomposition approach. *Eng. Optim.*, 48, 2141–2156, 2016.

21. Mansouri, M.M. and Javidi, Cost-based job scheduling strategy in cloud computing environments. *Distrib. Parallel Database*, 38, 365–400, 2019.

22. Makori, E.O., Promoting innovation and application of Internet of Things in academic and research information organizations. *Libr. Rev.*, 66, 8/9, 655–678, 2017.

23. Meena, O.P. and Somkuwar, A., Comparative analysis of information fusion techniques for cooperative spectrum sensing in cognitive radio networks. *Proceedings of International Conference on Recent Trends in Information, Telecommunication and Computing*, 2014.

24. Hoang, D.C., Real-time implementation of a harmony search algorithm based clustering protocol for energy-efficient wireless sensor networks. *IEEE Trans. Ind. Inf.*, 10, 774–783, 2014.

25. Abreu, D.P., Velasquez, K., Curado, M., Monteiro, E., A comparative analysis of simulators for the Cloud to Fog continuum. *Simul. Modell. Pract. Theory*, 101, 1–27, 2020.

26. Dey, A.S., Ashour, F., Shi, S.J., Fong, R.S., Sherratt, S., Developing residential wireless sensor networks for ECG healthcare monitoring. *IEEE Trans. Consum. Electron.*, 63, 4, 442–449, 2017.

27. Prema, G. and Narmatha, D., Performance of energy a ware cooperative spectrum sensing algorithm in cognitive wireless sensor network. *International*

Conference on Green Engineering and Technologies (IC-GET), Coimbatore, 19 November, 2016, 2016.

28. Gharaei, N., AbuBakar, K., MohdHashim, S.Z., Hosseingholi Pourasl, A., Siraj, M., Darwish, T., An energy-efficient mobile sink-based unequal clustering mechanism for WSNs. *Sensors(Basel)*, 17, 1–20, 2017.

29. Akyildiz, I.F., Lo, B.F., Balakrishnan, R., Cooperative spectrum sensing in cognitive radio networks: A survey. *Phys. Commun.*, 44, 40–62, 2011.

30. Barga, R., Gannon, D., Reed, D., The client and the cloud: Democratizing research computing. *IEEE Internet Comput.*, 15, 1, 72–75, 2011.

31. Mell, P. and Grance, T., The NIST definition of cloud computing, NIST Special Publication 800-145, 2011.

32. Youseff, L., Butrico, M., Da Silva, D., Toward a unified ontology of cloud computing, in: *Grid Computing Environments Workshop. IEEE*, pp. 1–10, 2008.

33. Parkhill, D., *The challenge of the computer utility*, Addison Wesley, Reading, Boston, USA, 1966, 2017.

34. Mizani, M.A., Cloud-based computing, in: *Key Advances in Clinical Informatics*, pp. 239–255, Academic Press, Cambridge, Massachusetts, USA, 2017.

35. Mastelic, T., Oleksiak, A., Claussen, H., Brandic, I., Pierson, J.M., Vasilakos, A.V., Cloud computing: Survey on energy efficiency. *ACM Comput. Surv.*, 47, 2, 33, 2015.

36. Buyya, R., Yeo, C.S., Venugopal, S., Broberg, J., Brandic, I., Cloud computing and emerging IT platforms: Vision, hype, and reality for delivering computing as the 5th utility. *Future Gener. Comput. Syst.*, 25, 6, 599–616, 2009.

37. Ali, M., Khan, S.U., Vasilakos, A.V., Security in cloud computing: Opportunities and challenges. *Inf. Sci.*, 305, 357–383, 2015.

38. Mansouri, N., Ghafari, R., Zade, B.M.H., Cloud computing simulators: A comprehensive review. *Simul. Modell. Pract. Theory*, 104, 1–50, 2020.

39. Mansouri, M.M. and Javidi, A review of data replication based on meta-heuristics approach in cloud computing and data grid. *Soft Comput.*, 24, 14503–14530, 2020.

40. Perez, A., Karima, V.M., Curado, E., Monteiro, A comparative analysis of simulators for the Cloud to Fog continuum. *Simul. Model. Pract. Theory*, 101, 1–63, 2020.

41. Gupta, K. and Beri, R., Cloud computing: A survey on cloud simulation tools. *Int. J. Innov. Res. Sci. Technol.*, 2, 11, 430–434, 2016.

42. Calheiros, R.N., Ranjan, C.A.F., De Rose, R., Buyya, H., CloudSim: A novel framework for modeling and simulation of cloud computing infrastructures and services, in: *Grid Computing and Distributed Systems Laboratory*, pp. 1–9, The University of Melbourne, Australia, 2009.

43. Bonomi, R., Milito, J., Zhu, S., Addepalli, F., Fog computing and its role in the Internet of Things, in: *Proceedings of the 2012 ACM First Edition of the MCC Workshop on Mobile Cloud Computing*, ACM, pp. 13–16, 2012.

44. Whaiduzzaman, A., Naveed, A., Gani, M., MobiCoRE: Mobile device based cloudlet resource enhancement for optimal task response. *IEEE Trans. Serv. Comput.*, 11, 144–154, 2016.

45. Chen, Y., Chen, Q., Cao, X., Yang, Y., PacketCloud: A cloudlet based open platform for in-network services. *IEEE Trans. Parallel Distrib. Syst.*, 27, 4, 1146–1159, 2016.

46. Yi, Z., Hao, Z., Qin, Q., Li, S., Fog computing: Platform and applications, in: *Proceedings of the 3rdWorkshop on Hot Topics in Web Systems and Technologies, HotWeb 2015*, Washington, DC, USA, pp. 73–78, 2016.

47. Dastjerdi, A.V., Gupta, H., Calheiros, R.N., Ghosh, S.K., Buyya, R., Fog computing: Principles, architectures, and applications. *Internet Things*, 1, 61–75, 2016.

48. Sarkar, S. and Misra, S., Theoretical modelling of fog computing: A green computing paradigm to support IoT applications. *IET Netw.*, 5, 2, 23–29, 2016.

49. Atzori, L., Iera, A., Morabito, G., The Internet of Things: A survey. *Comput. Networks*, 54, 15, 2787–2805, 2010.

50. Al-Fuqaha, A., Guizani, M., Mohammadi, M., Aledhari, M., Ayyash, M., Internet of Things: A survey on enabling technologies, protocols, and applications. *IEEE Commun. Surv. Tutorials*, 17, 4, 2347–2376, 2015.

51. Bandyopadhyay, D. and Sen, J., Internet of Things: Applications and challenges in technology and standardization. *Wirel. Pers. Commun.*, 58, 1, 49–69, 2011.

52. Khan, R., Khan, S.U., Zaheer, R., Khan, S., Future internet: The Internet of Things architecture, possible applications and key challenges. *2012 10th International Conference on Frontiers of Information Technology (FIT)*, IEEE, pp. 257–260, 2012.

53. Chen, S., Xu, H., Liu, D., Hu, B., Wang, H., A vision of IoT: Applications, challenges, and opportunities with china perspective. *IEEE Internet Things J.*, 1, 4, 349–359, 2014.

54. Li, S., Da Xu, L., Zhao, S., The Internet of Things: A survey. *Inf. Syst. Front.*, 17, 2, 243–259, 2015.

55. Lee, I. and Lee, K., The Internet of Things (IoT): Applications, investments, and challenges for enterprises. *Bus. Horiz.*, 58, 4, 431–440, 2015.

56. Gubbi, J., Buyya, R., Marusic, S., Palaniswami, M., Internet of Things (IoT): A vision, architectural elements, and future directions. *Future Gener. Comput. Syst.*, 29, 7, 1645–1660, 2013.

57. Perera, C., Zaslavsky, A., Christen, P., Georgakopoulos, D., Context aware computing for the Internet of Things: A survey. *IEEE Commun. Surv. Tutorials*, 16, 1, 414–454, 2014.

58. Lu, Y. and Da Xu, L., Internet of Things (IoT) cyber security research: A review of current research topics. *IEEE Internet Things J.*, 6, 2, 2103–2115, 2018.

10

Product Life Cycle

Harpreet Singh[1], Neetu Kaplas[2], Amant Sharma[3] and Sahil Raj[1]*

[1]*School of Management Studies, Punjabi University, Patiala, Punjab, India*
[2]*School of Humanities and Social Sciences, Thapar Institute of Engineering and
Technology (Deemed-to-be-University), Patiala, Punjab, India*
[3]*Independent Researcher, Patiala, Punjab, India*

Abstract

In this competitive era, organizations provide better consumer-centric products and services, improved market share and size, and consistently growing profits. Nevertheless, the global competition has increased the challenges in terms of prices, shorter lifecycle, more customized demand, complex products, the entrance of more suppliers, and strict government regulations; at present, these all have increased the numerous obstacles to manufacturers. Therefore, Product Lifecycle Management (PLM) has become a vital tool in getting a competitive advantage by creating and providing better quality products at less time and competitive price, sustainably achieving a compelling cost advantage. PLM supports the management to control the portfolio of the products and services, processes from the beginning of the concept, product designing, engineering, production, launch, and uses to disposal of the product. To cope with the global competition, organizations have to amalgamate PLM with technology such as Computer-Aided Design (CAD), Manufacturing Process Management (MPM), Computer-Aided Manufacturing (CAM), and Product Data Management (PDM); this systematic way helps in enabling the online sharing of product data as well as business applications. This coordinates and unifies the products, services, processes, and project data across the entire lifecycle among numerous external and internal players of an organization. This chapter discusses various aspects of the PLM, its process and workflows, PLM strategy and implementations, process modelling, modelling using UML, system architecture, Product data throughout the product lifecycle, Model-based systems engineering, and systems description techniques. These models help get insights about the application of PLM in the manufacturing industry and help PLM apply its concepts systematically for managing the

⋆Corresponding author: dr.sahilraj47@gmail.com

Chandan Deep Singh and Harleen Kaur (eds.) *Factories of the Future: Technological Advancements in the Manufacturing Industry*, (229–256) © 2023 Scrivener Publishing LLC

product-related information and processes across the value chain of a product. Such as Unified Modelling Language (UML) is a graphical modelling language used for specialized graphical presentation of business and system software; Product Data Management (PDM) directly deals with the centralizing data of a product; System Architecture is an abstract, global, conceptualized, and focused process in achieving the lifecycle concepts and mission of the system etc. PLM in the manufacturing industry connects the manufacturers and their stakeholders in a digital cloud-based environment; as a result, the manufacturer can run production much faster to meet customer demand timely.

Keywords: Product lifecycle management (PLM), unified modelling language (UML), product, process, strategies, planning, modelling, architecture

10.1 Introduction

Globalization has dramatically changed the operations of the manufacturing sector. On one side, globalization has offered many opportunities to attract new customers; as a result, it has increased intense competition, and enhanced global consumption. On the other side, this has also resulted in making organizational supply chains more critical and complicated [1]. The 2050 vision of the Convention on Biological Diversity and the United Nations Environment Program has estimated that global consumption will increase up to eight times and stressed the urgent need to focus on sustainable production and consumption on a national level [2]. This has put tremendous pressure on the manufacturing industry to improve efficiency in preparation, product development, manufacturing, and production planning [3]. To accomplish these objectives, manufacturing organizations are adopting the concept of Product Lifecycle Management (PLM) [1], which plays a key role in the smooth flow of goods and services from the point of producers to the point of end users along with the provision of after-sale services.

10.2 Product Lifecycle Management (PLM)

In the competitive world, organizations are morally bound to introduce environmentally sustainable fuel-efficient products, generate low noise, are bio-degradable, and cause less air pollution. Therefore, organizations need effective engineering to cater to this modern world, which can simultaneously be achieved by applying Product Lifecycle Management (PLM). This allows the organizations to maintain a competitive edge, especially at the initial product lifecycle stage, that comprises flexibility, productivity, and efficiency in a broad term [4]. PLM focuses on creating, modifying, and exchanging product-related information throughout the life cycle [6]. PLM systems use Computer-Aided Design (CAD), Manufacturing Process Management (MPM), Computer-Aided Manufacturing (CAM), and Product Data Management (PDM) [5, 6]. Hence, PLM is a consolidated technique that helps the management maintain the market share by providing updated product and market-related information and sustainable revenue for their product. PLM is an information-driven approach that entails continuous improvement, adding more innovative attributes, customer care services, etc. [7]. PLM provides seamless integration of complete information collected through all the stages of PLC and provides this information to the organization, supplier, and customer [12]. Product engineers can shrink this product cycle by implementing and selecting the changes across the supply chain.

10.2.1 Why Product Lifecycle Management?

Product Life Cycle Management is the extended version of the traditional Product Life Cycle. It also considers the data and related information generated at each stage of the product life cycle. This data and information are linked up to meet the requirement for the next stage of the product life cycle. The data generated at each stage of the product life cycle helps the company to retain the product positioning, and long-run survival of the product in the market as the product is manufactured and served according to the requirement and specifications of the target customers. In the current scenario, four stages of the old biological product lifecycle have been overcome by introducing the engineering field in product lifecycle management. This redefined Product life cycle management starts with concept generation. It passes through product design, raw material procurement, manufacturing, transportation, sales, utilization, and after-sales service and ends up with recycling/disposal of the product.

10.2.2 Biological Product Lifecycle Stages

The product life cycle is the journey of a product that starts from market acceptance and ends at elimination. The product lifecycle can be subdivided into four stages which have been discussed below:

Introduction Stage: The introduction stage begins with launching a product in the market to maximize the sale. This stage is termed as the "money sinkhole" stage as in this stage, organizations incur huge expenditures on advertising and promotion, and sales are lazy. The organizations also establish distribution channel that helps to spread the product on every counter of the market.

Another most crucial factor under this stage is pricing. Product pricing requires structured strategies according to the product; for instance, if the company is launching a different and unique product, the customer is ready to pay a high price, or an organization can follow pre-set price strategy for sale maximization. However, this requires sound knowledge of the market and customers' requirements. The market mix also plays a crucial role in determining the target audience before launching the product in the introduction stage.

Growth Stage: This stage shows a rapid increase in sales because early adopters and untargeted potential buyers start purchasing the product. During this phase, new competitors also enter the market, and as a result, they introduce new products with additional features. With an increase in the number of competitors, the customer demand for the product stabilizes, reducing the product's price.

To earn on investment and maintain the price, organizations check expenditures or slightly raise them to stay competitive. However, sales volume moves up faster many folds compared to its actual marketing expenditure. On the other side, an increase in market volume reduces the marketing expenditure and manufacturing cost, which helps to enhance the profit percentage. In that situation, firms must focus on the new strategies as the growth stage starts decelerating.

For maintaining sustainable growth, the organization should concentrate on:

- Improving the quality of a product while keep on adding new features,
- Entry into a new market (find a new market).
- Expand and adopt new methods of distribution in the existing as well as new markets.
- More focus should be on loyalty and preference rather than awareness and communication, in the sense that loyal customers should be given some additional benefits.
- Reduce prices to draw the attention of the new price-sensitive customer so that sales can be boosted.

Under this phase, the firm can get a dominant position by spending more on the improvement of product, advertisement and distribution channels to maximize the reach of the product. It escalates the product profit and market share, increasing the chances of profit maximization in the long run by consistently focusing on customer satisfaction.

Maturity Stage: After a certain period, sales volume goes down, and a product enters the maturity phase. The maturity stage can be bifurcated into three sub-stages: growth, stability, and decay. In the first stage, sales volume goes down. The firm avoids adding up a new distribution channel as new competitors move out. In the second stage, due to market saturation, per capita sales volume starts reducing because the potential customers have already used the product, and further, sale depends upon the growth of the population. In the last stage, sales start-up declining and consumer starts to look for a replacement. This phase is most challenging for the organizations; as a result, weak competitors move out. Big firms only survive because of cost, quality, and service leadership.

Declining Stage: In the final biological stage of a product life cycle, sales diminish due to various reasons such as a change in the taste and preference of consumers, technological advancement, and an increase in competition among domestic or foreign competitors. Under this situation, reducing the price through overcapacity utilization and profit erosion is the only solution available for the survival of the firms. The impact of this stage varies from industry to industry; therefore, many small or weak organizations withdraw. The big firms prefer to reduce the manufacturing of products, cutting down marketing expenditure and R&D costs, no investment in equipment, follow fewer trade channels, and a further price reduction. The stages mentioned above can be explained with the following example.

10.2.3   An Example Related to Stages in Product Lifecycle Management

A journey of the apple juice category in product lifecycle:

Introduction: As apple juice is available in all the stores with different kinds of packaging as per customer requirements and keeping this thing in mind marketing department takes all sorts of pricing and branding related decisions.

Growth: To make apple juice an eye-catching product, the marketer focuses on designing the packaging of apple juice bottles or containers. On the other hand, distributors work on increasing brand awareness and market shares.

Maturity: As per customer requirement, the packaging size and shape are changed, and calorie counts are added with the title like, "tastier and healthier". Marketer gains the customer's trust by providing a product that emphasizes diet rather than having a drink just for fun. Target to maximize the product sales is achieved.

Decline: Tough competition and changing demands lead to a downfall in sales which induces the company to move towards an exit strategy, even though the product has not been declined yet.

With the change in demand and zero tolerance in wastage, new phases and techniques have been adopted so that refined quality products at a due time can be delivered to the customers. A few new phases in the traditional Product Lifecycle Management have been introduced, discussed below.

10.2.4   Advanced Stages in Product Lifecycle Management

Concept generation: The idea of a new product is generated by gathering the information or tracking the customer demands, tastes, preferences, market trends, plans of investment, and some other sources too so that the product meets the requirement of all stakeholders can be produced.

Product design: At this stage, the product development team comes up with a product design with the help of exchanging ideas and concepts among the team. The data included in this stage consists of the product's appearance, applications of the product, data testing, and various other important components. The concept of 'Ergonomics' is considered as the most important factor at this stage.

Raw material procurement: At this stage, both manufacturers' and suppliers' data are recorded or analyzed. In manufacturers' data, performance, type, the quantity of raw material is given due importance. Important points from the suppliers' side such as costs and transportation distance are considered.

Manufacturing: It is a quality testing stage, and under this stage, the working of (Machines, performance, parameters, and conditions) is tested. The data is recorded in real-time, which proves beneficial for the organization in making managerial decisions.

Transportation: Keeping in view the market demands and approachability, the finished product is transported to the point of sales. Meanwhile, this step is carried out; the product is sold, and transportation services are provided to the end-users or buyers.

Sales: At this stage, it is considered that launching of product and carrying out marketing strategy totally based on the total order data, buyer's data, and suppliers data. Customers' preferences, tastes, convenience, crowd, and location are kept in mind, and product design and sales progress can be improved through concrete information.

Utilization: The information generated by the customer can be utilized in product improvement if required. The generated data is not limited to bringing some changes in the product or maintenance and repairing the existing product only. Still, it is also used to update the product design.

After-sales service: Maintenance of products and providing services is the main agenda. With the help of this data, proper or required services are provided to the customers, and if required, any sort of repairs can be done, and problems are sorted out. It results in the efficient working of services proved to customers.

Recycle/disposal: It is the last stage where the producer has to decide whether the product is worth discarding or whether it requires some recycling or repairing to be made sustainable in the market for a few months or years.

10.2.5   Strategies of Product Lifecycle Management

The entire product life cycle passes through four phases. Every phase has a unique characteristic, such as in the introductory phase requiring heavy advertisement, not on price, but in the growth phase, demands to increase the price; hence promotion expenditure declines. In contrast, both price and promotional expenditure tend to neutralize in the maturity phase. There is a decline in price and promotion [8]. Therefore, to keep the pace of a product during the life cycle, every organization makes numerous strategies as per product features and market requirements to maintain the smooth journey from introduction to final removal from market shelves. Following are strategies for product life cycle [9]:

- **Strategies applied in the introductory stage:** This stage takes a lot of time to gain the buyer's attention, resolve technical problems, and fill the distributor's gap. The organizations apply different strategies like advertising campaigns, spreading word of mouth through social media platforms, and promoting sales. Regardless of the slow sale and fewer profits, organizations set high prices due to increased production and promotional costs [10, 11]. Following are the strategies for the first stage of the Product Life Cycle.

10.3   High and Low-Level Skimming Strategies/Rapid or Slow Skimming Strategies

It is the most commonly used retail pricing method where a product is introduced in the market at a higher price, and when the competition starts increasing, then the company remains in survival or competition starts to reduce the prices so that demand for the product can be stagnated. We can further clarify this concept with the help of an example. For say, when a producer of company A, launches a mobile phone in the market, the prices are very high because of the craze and excitement to buy that product among the customers, but, steadily when the time passes, the prices get lower with the decrease of demand, and eventually, the company launches a new product. Just like many other big technology-related brands are doing their businesses in this segment. But this does not mean that the high and low-level pricing strategy is only limited to the technological sector; it is practiced in every field. This strategy is used for the small apparel companies/brands that set the prices of their new apparel range pretty high during the new seasons. With the changing season or in the middle of the season, the prices are set on discounts to increase sales, but top-notch brands never rarely use it.

10.3.1   Considerations in High and Low-Level Pricing

- In this type of pricing, once the buyer has bought the product at a lower price, there is no coming back as the seller cannot sell that same product at a high price.
- When there is an unexpected situation when the prices hit high (pandemic), this may negatively affect the business.
- It is advisable to observe customers' reactions while setting the prices high, and according to the reaction of the customers, the seller can set prices either high or low as per the situation demands.

10.3.2   Penetration Pricing Strategy

In this method of pricing strategy, the seller starts with selling the product at a low price to increase the sales and awareness among the customers. Still, the seller has already planned to increase the product's prices in the near future, unlikely in the skimming strategy.

10.3.3   Example for Penetration Pricing Strategy

When Mr. Mukesh Ambani launched Reliance Jio, India's one of the largest telecommunication network providers, it took the nation to the storm. As Jio had played an important role in the digital revolution and came up with a unique pricing strategy when Jio was launched, it did not charge any tariff. It had attracted the Indian people to their internet and telecommunication services without any surprise. Within a short span of time, Jio started to increase the prices for its telecom services; as a result, customers accepted their pricing policies. The reason behind accepting their pricing by customers was that Jio first made the customers habitual to avail of their tariff-free services. When they entirely became bound to avail of their internet-based services, the company started to increase the prices of their services slowly, which did not impact the customers' affordability.

10.3.4   Considerations in Penetration Pricing

- The businessman who is hustling to bring a revolution in the industry phase will face a bit of a hurdle because of these market penetrations.
- Setting the prices low at the starting point may help the business runner earn some goodwill and gain customers. But, it can be risky, too, as it is hard for the beginner to set the prices low at the start of a business.
- On a positive note, there is a breakeven analysis that helps the seller/businessman to know whether the chosen strategy is for them or not.

• **Strategies for growth stage:** Under this phase of rapid market expansion, an organization is focused on sustaining the market share for a more extended period. Here, a firm does not concentrate on promoting product awareness; instead, they put their energy into trialing the product. After a successful introduction in the domestic market, the firm tries to enter into new segments and expand the distributors' channel. Since the new competitors have already entered the market, the firm has also strengthened its competitive position.

The following are actions are considered under this stage:

- **Market penetration:** Sellers tend to attract their customers or keep them distracted from their (seller's) competitors and make sure to gain their trust. This will be achieved when the prices are low, and promotions of that product are high, basically to inform your customer that you are selling the best at a reasonable rate.
- **Market development:** when the product is already in the market, but the seller wants to increase its sales; this motive can be achieved when the sellers try in the new customer segments, changes the areas of sales, for say selling in a different country to increase the sales and awareness too.
- At some point, reduce the price to attract the price-sensitive customer by offering them with occasional discount.
- Preventing the competitors from entering the market by lowering the price of the product and increasing the promotional efforts.
- **Strategies for maturity stage:** Under this stage, the competitors that are already entered the market are juggling to get more market share; hence an organization follows aggressive market strategies to beat these competitors.

The following strategies can be considered suitable under this situation:

- **No new strategy:** Under this maturity phase, making no new strategy can be considered an effective strategy because sooner or later, there will be a decline in sales. Therefore, the marketer should try to save money by investing in a new profitable product.
- **Change the strategy to cater to the existing market:** This strategy is focused on increasing sales by increasing the number of usages of the existing product. The organization can survive either by attracting more customers or by expanding the use of a product.

Here are the following ways to induce new customers towards the existing product:

- Convincing the new customer to try the product
- Expanding in new market
- Attracting competitors' consumer base.

Strategy for product modifications:

- **Quality enhancement:** This involves safety improvement, reliability, speed, taste, efficiency, durability, etc. Modification of the product offers more satisfaction.
- **Feature enhancement:** This improves attributes such as weight, size, color, accessories, materials, get-up, etc. Attributes development leads to attractiveness, versatility, and convenience. Many organizations prefer product modification to sustain in the maturity stage.

Product modification provides benefits in many ways, such as:

- Product modification helps to promote the organization as a leading, dynamic and progressive organization.
- Product modification attracts loyal customers from a specific segment of the market.
- Product improvement is a small expense for an organization, but it provides many benefits to all stakeholders.
- Changes in attributes boost up the sales force and network channels.

Marketing strategies for decline stage: An organization formulates numerous strategies under the decline stage to sustain itself in the market. The first and most important task is to find critical weak features in the existing product. Firms usually formulate a committee known as Product Review Committee to detect it. This committee collects data from internal and external sources to evaluate the product. After analyzing the data, the committee submits a report, which helps to take corrective actions.

Following strategies (actions) are suggested:

- **Continue with the existing product:** This strategy is continued assuming that competitors will leave the market. The selling and promotion expenditure of the organization has already been reduced. Therefore, many organizations continue with an original product in the existing market segment where they are in profit and exit the rest of the segments.
- **Follow product enhancement:** Continuous improvement in the qualities and attributes helps push up sales, and the minor changes may also attract new customers.

- **Discontinue the product:** When the sales go down, and an organization cannot continue with the existing product even with product improvement, in such a situation, the organization decides to discontinue the product line.

10.4   How Do Product Lifecycle Management Work?

Effective PLM systems provide access to critical data to engineers. The core benefit of PLM, it allows flexibility from supply chain to product delivery. Product lifecycle management is a holistic approach, like other projects. It has pushed up the demand for a hi-tech manufacturing environment. And the requirement for PLM has increased in the manufacturing sector as products are becoming more complex and require continuous maintenance, even after their launch in the market.

PLM is a combination of product lifecycle and different system software. In the innovative era, every project demands some type of utility software because there are many steps in creating and marketing a product that involves various departments' employees and sometimes other industries. Some firms outsource some manufacturing process components, such as designing, etc.

Therefore, to maintain the smooth functioning of the product lifecycle, software becomes essential for an organization that comprises Product Data Management, Computer-Aided Design, Enterprises Resource Planning, Process modelling, Modelling using UML and system architecture, etc. The systems software puts together all the diverse components. It makes the process more straightforward, sorted, and efficient by providing a common platform for all the multiple teams to access all types of elements and perform their jobs effectively.

The foremost advantage of PLM software is efficiency. Its every aspect is managed by centralized software and simplifies the process on every step. It has the following advantages:

- The primary focus of PLM is to reduce the overall cost and process time of a supply chain from idea generation to delivery in the market. PLM shows the big picture to identify where the expenditure can bring down 0without compromising the quality.
- PLM provides a centralized communication system that reduces communication errors.

- Another benefit of PLM software is that it provides a centralized electronic platform to keep the communication history, documents, and data to reduce the paperwork. This makes it easy for an organization to check the project details and ensure that the product follows all the requirements.

The important part of PLM in product development is that it describes the implementation of tools and software, which helps track and control the product data. The data tracking comprises the technical identification of a product, identification of the producer, and types of material required to use. Therefore to elaborate the primary process of PLM strategy, UML is used. UML helps to translate the diagrams to evaluate the quantitative performance of PLM. Furthermore, it helps to detect causes of delay in the system.

10.5   Application Process of Product Lifecycle Management (PLM)

Application of **Product Lifecycle Management (PLM)** framework Process varies from organization to organization according to their requirement and size. Hence, the following are the common steps to implement the PLM process:

1. **Define the PLM implementation goals:** PLM is a stem from Computer Integrated Manufacturing (CIM) and Engineering Data Management. Integration aims to eliminate existing barriers, generate new ideas and projects, make portfolios, and streamline the value-generating chains.
2. **Inspect the existence of PLM foundation:** The foundation of PLM is given by vigorous product structure. It helps in defining the structured relation among the components and modules of a product. Apart from this, it unites the product-related documents, information such as product structure (its requirement, functions), maintenance procedure and cost analysis, etc.)
3. **Process selection:** Models include proven organizational practices that are used as an initial point of process innovation. The process implemented is selected from the PLM process list while keeping in mind its objectives and expected benefits.
4. **Identify the maturity level of the process:** Here, at this stage, we compare the firm's current process with the

selected processes in the past to know the exact outcome as compared to the expected outcome.

5. **Customization of reference model:** Although all processes are targeted differently, each process should reflect a particular need to meet the required objective. And if any gap is observed, then corrective action is taken immediately to remove the gap.

Apart from the process selection, to make the production process user-friendly **Unified Modelling Language (UML)** is the technique which comprises; **Unified:** Unite or merge the organization's information system and technology to get the finest engineering practices, **Modelling:** Number of models are applied to explain the system, **Language:** Translating the diagrams into the purpose of communication [13].

Hence, Unified Modelling Language (UML) is a graphical modelling language used for specialized graphical presentation of business and system software. UML is the most versatile modelling technique because its information-rich representation includes constantly testing the model's consistency, analyzing and translating all models into many specialized ways such as Gantt charts, etc. These days many modern machines are usually controlled by applying complex software models using UML. Therefore, this approach is supported by non-software systems. Under product modelling, various people are involved at different stages of the product lifecycle process. At the same time, UML favors communication among the different designers, process planners, project management and client, etc., which helps in getting a purpose-oriented view [14]. Nowadays, every customer has a unique taste, and customer demands customized products; hence, many organizations concentrate on high safety and quality products. So, to meet customer demands, the firms first look into users' shared data with the help of Product Data Management (PDM), which helps them analyze users' needs and requirements. The main focus of PDM remains on a product idea, design, and use. UML assists in providing various ways to evaluate the diagrams for more simplification. Still, it depends on the type of data the organization must utilize to get the desired results.

10.6   Role of Unified Modelling Language (UML)

Since the importance of strategic value has increased, the industry has started looking for ways to automate the various production software, keeping in mind how to downsize the overall cost and increase the overall efficiency of the production process. These techniques encompass visual programming, component technology frameworks, and pattern. Organizations look for strategies to manage the system complexity [15]. UML helps to resolve the following requirements:

1. The user gets ready-made visual modelling which further helps in developing and exchanging the model to simplify the production process.
2. It helps in understanding the modelling language by making it more simper.
3. UML also assists the organization with higher-level development in collaboration, components pattern, and framework.
4. Provides specialization and extensibility mechanism to get a better understanding of core concepts.
5. It is difficult for a business person to understand coding; therefore, UML is emerging as an essential tool to meet the demand, process, and functionality.

10.6.1   UML Activity Diagrams

UML refers to a standard technique that visualizes the designed system, and it is similar to the blueprints used in engineering fields. UML draws up the behavior and structure of the system. It is a visual language rather than a programming language. The diagram depicts the specific view of the model because each model is complex and hard to present in a single diagram. UML gives the option of multiple diagrams so that organizations can concentrate independently on multiple aspects of a system [13].

In UML, diagrams are further supplemented with text and other descriptions [16]. UML diagrams have been classified into two broad categories that are discussed below:

1. **Structural diagrams:** This focuses on the static and structural sides of a system. It includes a class, component, and object diagram [16].
2. **Class diagram:** It shows the presence of several classes, interfaces, attributes, and methods [13]. Class diagrams also provide help in recognizing the relationship between classes and objects.
3. **Object diagram:** This diagram presents the set of objects and their relationships with each object [13].

4. **Component diagram:** This shows the internal structure of the class and its interaction area with other parts of the system while using the parts, connectors, and ports. It is also similar to the class diagram, the only difference; it focuses on the individual part rather than detailing the entire class.

5. **Package diagram:** A package diagram represents the dependencies among the multiple packages along with their internal composition. Packages usually are used to organize UML diagrams into groups and make them understandable. However, its primary function is to set up the class in standard form and use case diagrams.

6. **Behavior diagram:** This diagram focuses on the dynamic side and behavior of a system. It involves the use of a case diagram, activity diagram, and state diagram [16].

7. **State-chart diagram:** Also known as State Machine Diagram, this diagram shows the condition or part of a system, as there are infinite instances responsible for changes in external stimuli during a specific period [13, 16].

8. **Activity diagram:** This diagram focuses on the flow control due to the happening of a particular event [13].

9. **Use case diagram:** It depicts a system's functionality, requirements, and interaction with the external agents. A use case represents the different conditions where the system is used.

10. **Communication diagrams:** A communication diagram or collaboration diagram is used to represent the sequences of exchanged messages with different objects.

10.7 Management of Product Information Throughout the Entire Product Lifecycle

Product Lifecycle Management (PLM) has two roots-first is Enterprise Management (further subdivided into Customer Relationship Management, Supply Chain Management, and material resource planning). The second is Managing the Product Information throughout the entire lifecycle of a product. For data management, organizations prefer virtual enterprises such as Product Data Management (PDM) and Computer-Aided Design or Computer Management Design (CAD/CMD) [17]. CAD was originated in 1980 to develop Geometric Models, but such designs are easily reused and manipulated. Consequently, Product Data Management systems came to light in 1980 [18]. PDM is an easy, secure, and quick data generator throughout the product lifecycle. In the first generation, the PDM system works effectively for the engineering activities, besides other activities such as Supply Chain Management, sales, marketing, and other external activities of the organizations. On the other side, modern web-based PDM has brought more accessibility among the manufacturing concerns. Hence, PDM is more flexible in generating data on various product lifecycle stages [17].

10.8   PDM System in an Organization

PDM offers the various functionalities as per the target market requirement on the basic level, such as data storage, revision, change or access control, document management, workflow management, and project management. The production management department faces many challenges in maintaining the inventory, preparing effective production plans, etc. To address these issues Enterprise Resource Planning (ERP) system has come into operation. Therefore, many manufacturing organizations prefer to implement PDM and ERP systems together [19].

10.8.1   Benefits of PDM

The key benefits of PDM have been discussed below:

1. PDM helps to find the correct information quickly
2. It improves productivity level and reduces cycle time
3. Assists in reducing error and various costs
4. Optimize the production resources
5. Helps to meet business as well as regulatory requirements

10.8.2   How Does the PDM Work?

Product Data Management system is a framework that provides the manufacturers with maintaining and controlling their engineered information, especially information related to new product development and engineering process. From a product view, it helps to arrange design-related concepts and retrieve information and other product-related data. From a process perspective, PDM manages procedural events, such as reviewing design, product release, etc.

10.8.3 The Services of Product Data Management

Product Data Management (PDM) directly deals with the central data of a product. Hence, this area works as the main information allocator to other functional departments. According to [20]; the functions of PDM have been divided into three categories, which are discussed below:

1. **Product planning:** This function involves collecting, selecting, and evaluating new ideas along with an integrated data catalog and managing the product requirement.
2. **Product structuring:** At this stage, the organization records material masters (material-related data) according to predefined schemas to minimize efforts to collect and reuse information.
3. **Configuration management:** It deals with the change management and configuration of the entire product lifecycle, which is helpful in linking up the whole process of all stages of the Product Life cycle.

   Apart from the services mentioned above, PDM helps perform a number of functions in the creation and updating of information, such as:

1. **Production planning:** It involves plant process generation and accessibility to all resources related to the entire machinery. It helps reduce process time and timely access to all resources during the production process.
2. **Sourcing:** Under this function, data related to standard items, suppliers' information for requesting quotations is retrieved from e-Source functionalities, which helps get raw material on time to avoid delays in the production process.
3. **Quality management:** Various quality methods such as Failure Mode and Effect Analysis (FEMA), creating control plans, and managing through quality inspections are incorporated, which is very helpful in making the product competitive in the market.
4. **Environment management:** It comprises managing and handling hazardous items, like scrap and recyclable items, to avoid casualties and optimum utilization of organizational resources during the production process. Hence safety of workers and optimum utilization of scarce resources should be the prime focus of the organization.

Even though systems like ERP, CAD, and PDM are involved in the decision-making process, other multiple software is also a part of the systems that support the whole supply chain and play a key role in generating isolated parts of data. Therefore to navigate the complex process of PLM, a system architecture is used to analyze the framework step by step that also covers the upcoming challenges. System architecture integrates the PLM system to visualize the information in a single supply chain atmosphere.

10.9   System Architecture

A system architecture is a framework that develops a common practice for establishing, analyzing, and representing the manufacturing system throughout PLM's entire designing and redesign process [17]. This is an abstract, global, conceptualized, and focused process in achieving the lifecycle concepts and mission of the system. The Functionality of the PLM system includes; Information Technology Infrastructure, Product Modelling Architecture, and another set of business applications. However, Information Technology is a foundation of PLM that encompasses software, hardware, and other internet-based technology, that helps to maintain the product and its related information across the product life cycle [21].

The highly dynamic current situation triggers the manufacturing system design; as a result, organizations are changing their design solutions at various dimensions and stages. The objective of the system architecture is to create an all-inclusive solution based on major components, such as principles, properties, and concepts, and that all are logically consistent with each other. However, each system design can be applied to more than one system, where patterns and requirements are homogeneous like related systems [22, 23].

Categorization of Principles and Heuristics Approach

A crucial step in the manufacturing system design process is getting the physical solutions with the operational needs and depicting system capabilities, supports, and design processes [22]. Therefore, architects, engineers, and designers use a combination of mathematical proven principles and heuristics approaches (learning via experiences, but these lessons are not mathematically proven).

These principles and heuristic approaches are used in the system that has further categorized into four domains:

1. **Static domain:** This is related to the system's physical structure; furthermore, organizations can be divided into systems and their elements.

2. **Dynamic domain:** It consists of the logical part of the architecture, especially the behavior of the system. It transforms input flow from output flow and interaction between the systems and external objects. Hence, the main objective of this domain is to maintain the effectiveness of the system.

3. **Temporal domain:** It deals with the execution of the function in a cyclic and synchronous way so that the effectiveness of the system can be observed.

4. **Environmental domain:** This domain deals with the system's survival during natural hazards or internal threats to the system's integrity and makes sure that the system is still operative even during the time of emergency.

10.9.1 Process of System Architecture

The objective of system architecture is to build the various alternatives of the system architecture and select the best alternative to meet all requirements of stakeholders within a consistent view [22, 24]. The following process is adopted for system architecture:-

1. **Defining the stakeholder's requirement:**
   The initial step of system architecture is to understand the context for which purpose the system is required to use according to the stakeholder's requirement or concern. For that, engineers need to study the industry, market, operations, mission and vision, legal, and other related information that may help in getting a better understanding of building the system architecture.

2. **Identify the necessary system viewpoints:**
   Identifying the relevant viewpoints of various stakeholders and frameworks to support the development models is considered.

3. **Develop models of architecture:**
   Define the entities to support the architecture, such as physical interface, functions, input/output, system and information elements, nodes, links, etc. It helps meet the stakeholder requirements (functional, interface, operational, human factor, etc.

In this stage, models are verified by implementing and checking the architecture's validity, reliability, and feasibility.

4. **Relate system architecture and design:**
   Define the system elements that represent the architectural features. For this, align, partition, and allocate system requirements to its elements to satisfy the architectural characteristics.

5. **Establish and maintain the selected architecture:**
   In the end, develop a means for the system architecture's governance (including roles, authorities, responsibility, and control functions). Manage the evolution and maintenance of the architecture procedure; this comprises; complete changes as per environment, technological context, operational experience, and implementation. Also, verify the impacts of the governance on architecture during the lifecycle. Finally, coordinate the reviews of architecture to achieve the user agreement.

Some Key Points for an Efficient System Design

Architecture is a blueprint of the system. It manages the complexity of a system and generates coordination and communication between the components and their users. The system design is outlined as scalable, efficient, cost-friendly, and secure architecture [15].

Following are the key factors for an effective system architecture:

1. **Consistent and integrity:**
   Before designing a reliable and error-free architecture, the highest consideration part is to focus on the reliability and integrity of the system's data. The system must have redundant options of backup to maintain the data consistently.

2. **Security:**
   In the modern hi-tech era, universal computing has made the users' data-sensitive, and the overall information security of the system is supreme. The system architecture must incorporate a strong security policy through configuration.

3. **User friendly:**
   The user-friendly architecture is the foremost part of the system design. The overall process of users' journey should be articulate from one part to another part of the system, and this should also include proper indications and navigation to provide them with better insights into the application.

4. **Planning and recovery:**
Data planning and recovery are crucial parts of the system design that check out the organization's requirements and should not be interrupted by unforeseen events. These unforeseen and unwanted events can be handled efficiently with strong system planning.

5. **Continuous testing:**
Unit testing is a resilient feature of system architecture. It is built up to generate the report to review where any irregularity can find be quick, which helps take the immediate step to control the irregularities.

The Model-Based System Engineering (**MBSE**) approach plays an important role in supporting the system architecture. It is a formalized framework that supports the specification, analysis, design, validation, and verification of the complicated system. MBSE keeps the model in the middle of system design that provides an effective way to update, explore and transmit the system facets to the users while removing and decreasing dependence on traditional methods [25, 26]. MBSE is an effective way to explore and document the system attributes. While continuously testing and ratifying the characteristics and timely learning the properties of a system helps to speed up the feedback process and its design decisions.

10.10 Concepts of Model-Based System Engineering (MBSE)

Model-Based System Engineering (MBSE) combines the three different concepts which have been discussed below:

1. **Model:** A model is an understandable version of a mathematical, graphical, and physical depiction that helps build a reality after eliminating some complications. A system architect should design the system with fewer details to maintain the structure easily. In short, a model should depict the system sufficiently, and the system should verify the models.

2. **System thinking:** In the system thinking stage, the engineers explore its boundaries, lifecycle, and content. It further helps to identify missing interaction, duplication of work, missing steps in the process, etc. and guides system complexity

management. Here systems are more emphasized in their interconnection with other subsystems.

3. **Systems engineering:** It is an integrated and interdisciplinary approach that includes technological, scientific, and management methods. It is a bundle of techniques that makes sure that all the conditions are fulfilled as per the design system. Its main objective is to focus on architecture, integration, analysis, management, and implementation of the system successfully in its lifecycle. Hardware, software, processes, and methods of the system are also considered.

10.10.1   Benefits of Model-Based System Engineering (MBSE)

MBSE has multi-aspect and multidisciplinary endeavors. It is a structured application to support system design, specification, analysis, validation, and verification of various activities in the initial conceptual design phase. It remains continued during the entire development and lifecycle stages.

The key benefits of MBSE have been discussed below:

a.   It formalizes the procedure of systems development over its use to be made easy to operate.

b.   Broaden ups the scope by adding multiple modelling domains throughout the lifecycle; hence it tries to cover all aspects of the system to make the production process more convenient.

c.   Improves the quality and productivity and also reduces the overall risk

d.   It makes communication easy and strong between all the stakeholders

e.   It manages the system complexity

10.11   Challenges of Post-COVID 19 in Manufacturing Sector

The consequences of unusual challenges of the COVID-19 pandemic have been observed in the diverse manufacturing sectors across all the nations. Negative impact on the global economy and unexpected and quick changes in demand and supply have put all the businesses in an unstable situation [27]. During the pandemic period, many organizations took extreme steps to survive. For instance, Harley Davidson-reduced its non-essential expenditure, temporary CEO and board of directors forgoes their salary and other cash compensation, 10 to 20%

decrease in salary of other employees, and freeze up the new hiring, etc. New challenges in the manufacturing industry are ensuring a safe and healthy environment for their employees to cover up the rapid increase in customers' demand. Pandemic has increased the demand for Personal Protective Equipment (PPE), World Health Organization has recommended increasing the production of PPE by 40%. To meet the demand and supply of medical supplies across the globe, many communities around the globe have come together, such as Open Source COVID-19 medical supplies, etc., and for front-line workers, the project was Project N95.

To control the COVID-19, every nation has taken actions such as flight bans, slowdown the import, and export, and checking on other outside activities. In the manufacturing sector, risks have escalated, which contains an interrupted supply chain because of the slowdown in raw material production and spare parts and the uncertainty for recovery [28].

10.12   Recent Updates in Product Life Cycle

Recently, a big transformation has been observed in the Product Life Cycle, which is clear from the technology-driven business like Amazon and Uber. These data-driven businesses have raised the expectations of the customers. The networks emerge where data is collected from different Product Life Cycle phases. This data is utilized to handle the various issues arising during the design, manufacturing, usage after marketing, recycling, and remanufacturing of returned products.

In 2022, the database will be on par with the people, technology, and capital as a godsend for corporations. The negligence of not taking care of concerning data and deputing the data-centric operations to the third logistics providers (3PL) is becoming the chronology. The obstacle in the remanufacturing sector can be data residing with third-party logistics (3PL). For most companies' Original Equipment Manufacturing (OEM) based on data specific is in fashion these days as competition is rising and monitoring the defects and getting the feedback from customers to remove these defects has become a very crucial part of the business and manufacturing process in Product Life Cycle. Various internet-based technologies such as the Internet of Things (IoT), sensors, and other tools are being used to collect data of potential customers to deliver quality services at less cost and in a more sustainable way are the companies' prime motives. By adopting a digital manufacturing system following issues can be solved during the Product Life Cycle.

Customers Satisfaction: Digitization of manufacturing can be very useful in handling the customers' grievances in a very effective way; for example, if the customers return any electronic product that they purchased through the online platform, by giving them instant feedback to the manufacturer which was collected from the customer, will help the manufacturer to take immediate action on the issues emerged with the product. Major telecom companies in the US are adopting such changes.

Cost Reduction: Some telecom companies in the US have adopted the standard practice for remanufacturing the product within their warranty framework by following the feedback from the retailers that why that product was returned. This has helped the companies reduce their costs by 30 to 40 % by cutting the remanufacturing steps.

Development of New Business Models: The rise in customers' demand has motivated the companies to adopt the digital and automated manufacturing system based on the data collected from the customers. The data collected from the customers helps the companies and stakeholders to modify the product according to the customers' demands. This data is further sold out to the other stakeholders who are directly or indirectly connected with the Product supply chain for their use. **For example,** Amazon sells the data it collects from the customers to other stakeholders, and hence it monetizes that data to generate income for Amazon.

Digitization and automation of the manufacturing process played a key role in Product Life Cycle Process transformation, leading to customer retention for a long time with the company and sustainable business. Quick response to customer feedback will decide the companies' future growth based on digitization.

10.13   Conclusion

Why do people buy organic food? The moderating role of environmental concerns and trust. In this era of globalization, Competition among companies and high consumer expectations are the key drivers that have forced companies to change their traditional methods of production to modern methods of production. In the competitive world, organizations are morally bound to introduce environmentally sustainable fuel-efficient products, generate low noise, are biodegradable, and cause less air pollution. Therefore, organizations need effective engineering to cater to this modern world, which can be achieved by applying Product Lifecycle Management (PLM). Product Life Cycle Management is the extended version of the traditional Product Life Cycle. PLM considers the data-related information generated at each stage of the product life cycle, and it focuses on creating, modifying, and exchanging product-related information throughout the life cycle. PLM systems use- Computer-Aided Design (CAD), Manufacturing Process Management (MPM), Computer-Aided Manufacturing (CAM), and Product Data Management (PDM). It is a consolidated technique that helps the management maintain the market share by providing updated product and market-related information and sustainable revenue for their product. PLM is an information-driven approach that entails continuous improvement, adding more innovative attributes, customer care services, etc. With the introduction of the engineering field in the traditional Product Life Cycle (PLC), the four traditional PLC stages have been redefined under Product life cycle management, starting with concept generation. After that, it passes through product design, raw material procurement, manufacturing, transportation, sales, utilization, and after-sales service, and ends up with recycling/disposal of the product. The data generated at each stage of the product life cycle helps the company retain the product positioning and long-run survival of the product in the market as the product is manufactured and served according to the requirement and specifications of the target customers.

References

1. Kim, G.Y., Do Noh, S., Rim, Y.H., Mun, J.H., XML-based concurrent and integrated ergonomic analysis in PLM. *Int. J. Adv. Manuf. Technol.*, 39, 9–10, 1045–1060, 2008.
2. Lucas, P.L., Kok, M.T., Nilsson, M., Alkemade, R., Integrating biodiversity and ecosystem services in the post-2015 development agenda: Goal structure, target areas and means of implementation. *Sustainability*, 6, 1, 193–216, 2014.
3. Kim, G.Y., Lee, J.Y., Park, Y.H., Noh, S.D., Product life cycle information and process analysis methodology: Integrated information and process analysis for product life cycle management. *Concurrent Eng.*, 20, 4, 257–274, 2012.
4. Stark, J., *21st Century paradigm for product realisation*, Springer, New York, 2015.
5. Demoly, F., Yan, X.T., Eynard, B., Gomes, S., Kiritsis, D., Integrated product relationships management: A model to enable concurrent product design and assembly sequence planning. *J. Eng. Des.*, 23, 7, 544–561, 2012.
6. Srinivasan, V., An integration framework for Product Lifecycle Management (PLM). *Computer-Aided Des.*, 43, 5, 464–478, 2011.

7. Sivam, S.S.S., Kumar, S.R., Karuppiah, S., Rajasekaran, A., Competitive study of engineering change process management in manufacturing industry using product life cycle management – A case study, in: *2017 International Conference on Inventive Computing and Informatics (ICICI)*, 2017, November, IEEE, pp. 76–81.

8. Kamthe, M. and Verma, D.S., Product life cycle and marketing management strategies. *Int. J. Eng. Res. Technol. (IJERT)*, 2, 4, 2035–2042, 2013.

9. Andriotis, K., Strategies on resort areas and their lifecycle stages. *Tour. Rev.*, 56, 40–43, 2001.

10. Kotler, P. and Keller, K.L., *A framework for marketing management*, p. 352, Pearson, Boston, MA, 2016.

11. Anderson, C.R. and Zeithaml, C.P., Stage of the product life cycle, business strategy, and business performance. *Acad. Manage. J.*, 27, 1, 5–24, 1984.

12. C.R. and Zeithaml, C.P., Stage of the product life cycle, business strategy, and business performance. *Acad. Manage. J.*, 27, 1, 5–24, 1984.

13. B. Eynard, T. Gallet, P. Nowak, and L. Roucoules, UML based specifications of PDM product structure and workflow, *Computers in Industry*, 55, 3, 301–316, 2004.

14. Thimm, G., Lee, S.G., Ma, Y.S., Towards unified modelling of product life-cycles. *Comput. Ind.*, 57, 4, 331–341, 2006.

15. Eriksson, H.E. and Penker, M., *Business modeling with UML*, pp. 1–12, Wiley, New York, 2000.

16. Pooley, R. and King, P., The unified modelling language and performance engineering. *IEE Proc.-Softw.*, 146, 1, 2–10, 1999.

17. Benkamoun, N., ElMaraghy, W., Huyet, A.L., Kouiss, K., Architecture framework for manufacturing system design. *Proc. CIRP*, 17, 88–93, 2014.

18. Ameri, F. and Dutta, D., Product Lifecycle Management (PLM): Closing the knowledge loops. *Comput.-Aided Des. Applic.*, 2, 5, 577–590, 2005.

19. Ou-Yang, C. and Chang, M.J., Developing an agent-based PDM/ERP collaboration system. *Int. J. Adv. Manuf. Technol.*, 30, 3–4, 369, 2006.

20. Schuh, G., Rozenfeld, H., Assmus, D., Zancul, E., Process-oriented framework to support PLM implementation. *Comput. Ind.*, 59, 2–3, 210–218, 2008.

21. Sudarsan, R., Fenves, S.J., Sriram, R.D., Wang, F., A product information modeling framework for Product Lifecycle Management (PLM). *Computer-Aided Des.*, 37, 13, 1399–1411, 2005.

22. Sheridan, T.B., *Humans and automation: System design and research issues*, Human Factors and Ergonomics Society, Boca Raton, FL, 2002.

23. Shen, W., Hao, Q., Yoon, H.J., Norrie, D.H., Applications of agent-based systems in intelligent manufacturing: An updated review. *Adv. Eng. Inf.*, 20, 4, 415–431, 2006.

24. Xuemin, Z., Zhiming, S., Ping, G., The process of information systems architecture development. *Proc. Eng.*, 29, 775–779, 2012.

25. Hart, L.E., Introduction to model-based system engineering (MBSE) and SysML, in: *Delaware Valley INCOSE Chapter Meeting*, 2015, July, vol. 30, Ramblewood Country Club, Mount Laurel, New Jersey.

26. Batarseh, O., McGinnis, L., Lorenz, J., 6.5. 2 MBSE supports manufacturing system design, in: *INCOSE International Symposium*, 2012, July, vol. 22, No. 1, pp. 850–860.

27. Diaz-Elsayed, N., Schoop, J., Morris, K.C., Realizing environmentally-conscious manufacturing in the post-COVID-19 era. *Syst. Integr.*, 2, 3762–3012, 2020.

28. Cai, M. and Luo, J., Influence of COVID-19 on manufacturing industry and corresponding countermeasures from supply chain perspective. *J. Shanghai Jiaotong Univ. (Sci.)*, 25, 4, 409–416, 2020.

11

Case Studies

Chandan Deep Singh[1*] and Harleen Kaur[2]

[1]Department of Mechanical Engineering, Punjabi University, Patiala, Punjab, India
[2]DELBREC Industries Pvt. Ltd., Chandigarh, India

Abstract

The main concentration of these case studies has been to search and analyze various aspects on which the working of the organization depends. It involves the problems faced by firms in the era of globalization and need for technology advancement and the role of manufacturing competency in improving their performance. The case studies reveal that the growth of companies has been taking place on a steady pace since the past years. The timely implementation of new strategies and technologies has been the underlying factor for this development. Besides, it is the need of the hour for companies to make innovation and competency significant factors in their development plan if they want to survive in the competitive market. As obtained from quantitative, qualitative and modelling techniques, the factors Sustainable Development, Green Manufacturing, Advanced Production Techniques (Computer Aided Manufacturing and Advanced Maintenance) have a considerable impact as compared to others. The present research attempts to come up to the expectations of industrialists, policy makers, academicians by evaluating the impact of production facilities, upon which the success of the industry depends. In this research high priority risk factors of industry are taken up so as to provide a suitable qualitative model based on these. Taking into account the literature survey, the need of the present study arose because impact of sustainable green development through advanced manufacturing and maintenance techniques in manufacturing performance of manufacturing industry has not been yet addressed. For this purpose, the present study has been designed to investigate and suggest the parameters that contribute to the success of manufacturing industry. Moreover, this topic is need of the hour as it involves environment. For this purpose, the present study is so designed to investigate the sustainable green development initiatives that contribute to the

---

*Corresponding author: chandandeep@pbi.ac.in

Chandan Deep Singh and Harleen Kaur (eds.) *Factories of the Future: Technological Advancements in the Manufacturing Industry*, (257–284) © 2023 Scrivener Publishing LLC

performance of the industry. Based on the risks involved in the industry case studies have been developed based on sustainable green development which will go a long way in serving industries as well as academicians. To summarize, this research makes a significant contribution in the direction of sustainable green development through advanced manufacturing and maintenance techniques. However, this study helps to overcome the limitations that were encountered with the most methodological sound techniques.

Keywords: Sustainable development, green manufacturing, advanced manufacturing technology, computer aided manufacturing, advanced maintenance techniques

11.1   Case Study in a Two-Wheeler Manufacturing Industry

This organization is the globally largest manufacturer of two wheelers. It started its operations in India in 2001 at Manesar (Haryana) and has acquired over 12 million customers in 12 years of operations. It is now recognized as the fastest growing two-wheeler company in India.

Two-wheeler segment is the most important among the automobile sector that has experienced significant changes over the years. The two-wheeler sector consists of three segments viz. motorcycles, mopeds and scooters. The key manufacturers in this sector are TVS, Yamaha, Hero and Bajaj. Governance structure of this manufacturing unit in shown in Figure 11.1.

11.1.1   Company Strategy

The principle that it follows, is followed by all its companies worldwide.

Company Principle (Mission Statement)
People here are committed to provide products of best quality and at reasonable prices in order to satisfy customers worldwide.

Corporate Governance Structure

Corporate Governance

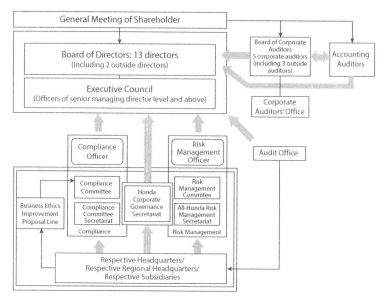

Figure 11.1 Corporate governance structure.

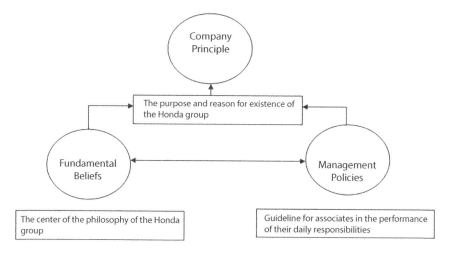

Figure 11.2 Its philosophy.

Fundamental Beliefs

Respect for the Individual

- **Equality**
  It means to respect and recognize differences in one another and fair treatment to everyone. It is dedicated to this principle, thus, creating equal opportunities for everyone. Figure 11.2 depicts its philosophy.
- **Initiative**
  It implies thinking creatively and acting on one's own judgment without being bound by preconceived notions and one must be responsible for the results from those actions.
- **Trust**
  Mutual trust is the basis for the relationship between associates here. Trust is created by helping each other, sharing knowledge, recognizing and making efforts for fulfilling one's responsibilities.

Management Policies

- Always have ambition and youthfulness.
- Making effective use of time and developing new ideas.
- Enjoying work and encouraging open communication.
- Consistently striving for a harmonious work flow.
- Be mindful of the value of endeavor and research. Figure 11.3 depicts principle initiatives in product development.

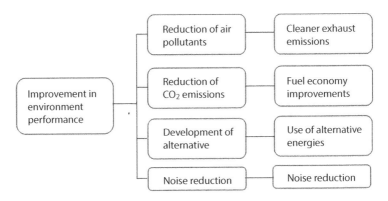

Figure 11.3 Principle initiatives in product development.

Innovation in Manufacturing: Strengthening the Fundamentals
To meet demand, it is pursuing innovations in manufacturing technology. They meet the demands and the expectations of customers and their stakeholders.

Vision of Environmental Technology
Implementing different technologies, it is delivering on the promise of genuine value in environmental responsibility and driving pleasure. Figures 11.4 and 11.5 shows principle initiatives in production and recycling respectively. Table 11.1 gives 3R for recycling.

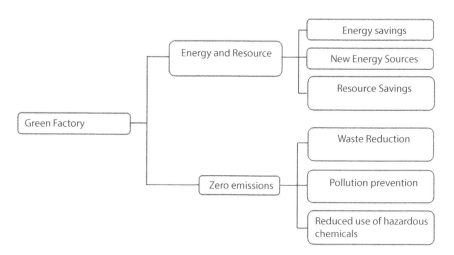

Figure 11.4 Principle initiatives in production.

Figure 11.5 Principle initiatives in recycling.

Table 11.1 Recycling (3R).

	Development	Products	Use	Disposal
Reduce	Design for reduction			
Reuse	Design for reusability & recyclability	Recycled/Reused Parts		
Recycling	Recycling and recovery of bumpers			Recycling of IMA batteries
		Recycling of by products		Compliance with the End of life vehicle recycling law in Japan
	Reduction in hazardous or toxic substances			Voluntary measures for recycling motorcycles

11.1.2   Initiatives Towards Technological Advancement

Cutting-Edge Technology

The fundamental design philosophy strives to maximize comfort and space for people, while minimizing the space for mechanical components. With this in mind, its R and D activities include fundamental research and product-specific development.

Combi Break System

Generally, it is quite difficult to control a two wheeler while braking during bad road conditions and emergencies. Combi break system allows easy and simultaneous operation of the rear and front brake and also providing optimal braking performance.

Matic Transmission

The efficient, compact and oil pressure controlled transmission is globally the first fully automatic transmission system, delivering a dynamic combination of torque and accelerator response for superior and constant driving experience.

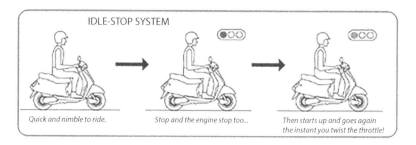

Figure 11.6 Idle-stop system.

Fuel Injection System

Its fuel injection system is designed to realize ideal combustion, resulting maximum power output, improving fuel efficiency and yet stay environment friendly.

Idle Stop System

It has developed an advanced Idle Stop System, as shown in Figure 11.6, which reduces fuel consumption and totally blocks out toxic exhaust gases and any unwanted noise. As soon as the vehicle stops, the engine is stopped automatically. And the engine restarts, when the throttle is opened and takes off smoothly.

11.1.3   Management Initiatives

1. *Respecting Independence*: (Challenge)
It expects associates to express their independence and individuality. In present time, associates are encouraged to think, act and accept responsibility. Anyone with proposals and ideas should express them.

2. *Ensuring Fairness*: (Equal Opportunity)
It offers a simple system with fair rewards for anyone having same abilities in handling similar sort of work and producing similar results, having no concern for nationality or race or gender, making no distinctions on educational basis or career history and objectively assessing each individual's strengths and aptitudes.

3. *Fostering Mutual Trust*: (Sincerity)
It believes that mutual respect and tolerance lays the foundation for trust that binds the employee and the company. Honda's personnel policy fundamentals are provided in Figure 11.7.

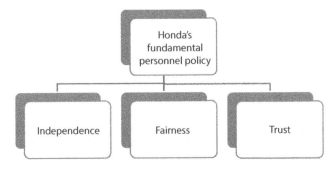

Figure 11.7 Principles of personnel management.

Fun Expansion

It is the first industry to promote green environment and safety in India. Since 5 years, It has expanded popular initiatives such as Asia Cup, One Make Race, Gymkhana and Racing Training by Moto GP riders from Japan.

Environment Conservation

On environmental front, it believes that tomorrow should be greener than today. For ensuring joy for next generation, it has implemented environmental management at its premises. It makes various efforts like reusing

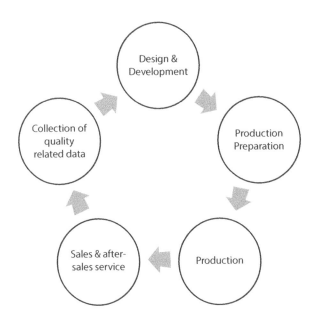

Figure 11.8 Quality circle.

and reducing waste for achieving zero emissions, improved efficiency and promotion of Green Factory, Green Supplier, Green Dealer initiatives and resource conservation.

Quality Assurance

It established Quality Innovation centers so that quality issues do not arise and enhance the capacity to resolve problems whenever they arise. Specialized departments at these centers are fully equipped to handle cases globally. They provide the resolution of any quality issues, rapid information and timely diagnosis. They also keep technicians and customers to date by providing the updates on recommended maintenance techniques. Its quality circle is shown in Figure 11.8.

11.1.4   Sustainable Development Goals

Being world's leading manufacturer in two wheeler sector, it is working to increase fuel efficiency and lower emissions. People here are working hard to improve environmental performance. For this, Programmed Fuel Injection (PGM-FI) has been implemented. It adapts to changes in engine load caused due to acceleration and deceleration, driving conditions, adjusting the volume and timing of fuel injection as well as the timing of ignition for optimal electronic control. With this, fuel efficiency gets improved and emissions are reduced without compromising on performance.

The **Sustainable Development Goals (SDGs)** or **Global Goals** are a collection of 17 interlinked global goals designed to be a "blueprint to achieve a better and more sustainable future for all". **Five** critical dimensions: people, prosperity, planet, partnership and peace, also known as the **5P's**. However, it actually refers to **four** distinct areas: human, social, economic and environmental – known as the **four** pillars of **sustainability**. Human **sustainability** aims to maintain and improve the human capital in society. The SDGs aim to "ensure that all human beings can enjoy prosperous and fulfilling lives and that economic, social, and technological progress occurs in harmony with nature." Inequality is one of the defining issues of this generation and requires a commensurate **focus** that, to date, has been lacking. Various SDGs are:

1. No poverty
2. Zero hunger
3. Good health and well-being
4. Quality education
5. Gender equality

6. Clean water and sanitization
7. Affordable and clean energy
8. Decent work and economic growth
9. Industry, innovation and infrastructure
10. Reduced inequalities
11. Sustainable cities and communities
12. Responsible production and consumption
13. Climate action
14. Life below water
15. Life on land
16. Peace, justice and strong institution
17. Partnerships for the goals

These **17 objectives are interrelated** and often the key to one's success will involve the issues most frequently linked to another. They can be summarized as follows:

- Eradicate poverty and hunger, guaranteeing a healthy life
- Universalize access to basic services such as water, sanitation and sustainable energy
- Support the generation of development opportunities through inclusive education and decent work
- Foster innovation and resilient infrastructure, creating communities and cities able to produce and consume sustainably
- Reduce inequality in the world, especially that concerning gender
- Care for the environment combating climate change and protecting the oceans and land ecosystems
- Promote collaboration between different social agents to create an environment of peace and sustainable development.

Develop sustainable, resilient and inclusive infrastructures; promote inclusive and sustainable industrialization; increase access to financial services and markets; upgrade all industries and infrastructures for sustainability; enhance research and upgrade industrial technologies. The remaining three targets are "means of achieving" targets: Facilitate sustainable infrastructure development for developing countries; support domestic technology development and industrial diversification; universal access to information and communications technology.

Implement the 10-Year Framework of Programs on Sustainable Consumption and Production Patterns; achieve the sustainable management and efficient use of natural resources; reducing by half the per capita global food waste at the retail and consumer levels; achieving the environmentally sound management of chemicals and all wastes throughout their life cycle; reducing waste generation through prevention, reduction, recycling and reuse; encourage companies to adopt sustainable practices; promote public procurement practices that are sustainable; and ensure that people everywhere have the relevant information and awareness for sustainable development. The three "means of achieving" targets are: support developing countries to strengthen their scientific and technological capacity; develop and implement tools to monitor sustainable development impacts; and remove market distortions, like fossil-fuel subsidies, that encourage wasteful consumption.

Some of the goals compete with each other. For example, seeking high levels of quantitative GDP growth can make it difficult to attain ecological, inequality reduction, and sustainability objectives. Similarly, increasing employment and wages can work against reducing the cost of living. On the other hand, nearly all stakeholders engaged in negotiations to develop the SDGs agreed that the high number of 17 goals was justified because the agenda they address is all-encompassing. Environmental constraints and planetary boundaries are underrepresented within the SDGs. For instance, the paper "Making the Sustainable Development Goals Consistent with Sustainability" points out that the way the current SDGs are structured leads to a negative correlation between environmental sustainability and SDGs. This means, as the environmental sustainability side of the SDGs is underrepresented, the resource security for all, particularly for lower-income populations, is put at risk. This is not a criticism of the SDGs per se, but recognition that their environmental conditions are still weak. The SDGs have been criticized for their inability to protect biodiversity. They could unintentionally promote environmental destruction in the name of sustainable development.

Several years after the launch of the SDGs, growing voices called for more emphasis on the need for technology and internet connectivity within the goals. In September 2020, the UN Broadband Commission for Sustainable Development called for digital connectivity to be established as a "foundational pillar" for achieving all the SDGs. In a document titled "Global Goal of Universal Connectivity Manifesto", the Broadband Commission said: "As we define the 'new normal' for our post-COVID world, leaving no one behind means leaving no one offline".

Despite organizations adopting Sustainable Development Goals, Organization growth has improved over the years so leading to an improved profit.

Figure 11.9 shows operating revenue for last 5 years of this two-wheeler manufacturing unit. The operating revenue has increased consistently, even though the industries are working towards sustainable development along with environmental issues.

The graph in Figure 11.10 shows the total income in the last five financial years. A gradual rise in total income has been witnessed during this phase. Total income is the sum of the money received, including income from services or employment, payments from pension plans, revenue from sales, or other sources.

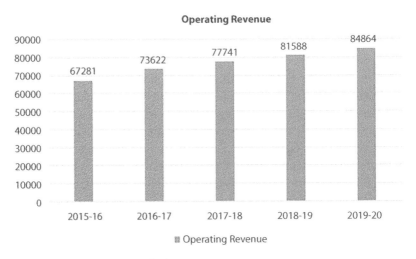

Figure 11.9 Operating revenue for last 5 years.

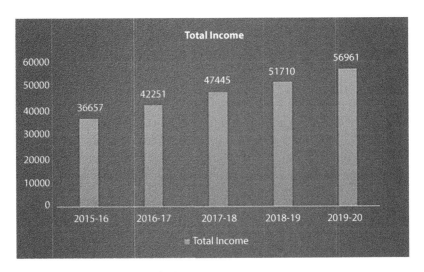

Figure 11.10 Total income of IT for last 5 years.

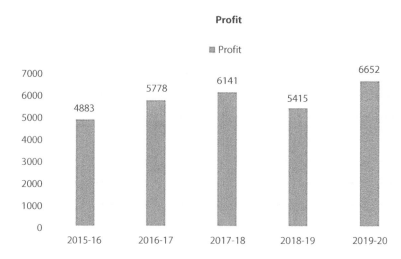

Figure 11.11 Profit for last 5 years.

Figure 11.11 shows the profit for the last 5 years. Even though there is a dip in FY'19 (2018-19) and there's a rise during last year.

From the above data, it has been observed that the company has been growing since the past five years. This is due to the company introducing new strategies and technologies in their products. It is very difficult to withstand the competitive world of the market otherwise. Every company has to make use of competency in their products in the present times. If no innovation is made in the product, it becomes obsolete. *It has legacy of cutting-edge Research and Development resulting in customer-oriented products. Today, due to the new technical centre, India is the centre of attention worldwide. They are devoted for delivering the best quality products at reasonable prices and at faster speed by having Research and Development, engineering, purchasing, quality and designing at same place.*

11.1.5   Growth Framework with Customer Needs

1. Premises/Process
- *Voice-of-the-Customer at dealerships*: evaluating customer feedback and bring it on operations.
- *Process efficiency improvement*: Improve work efficiency by elimination of wasteful operations at individual dealerships.
- *Single repair programs*: Ensure that customer's most issues are solved in a single repair. Figure 11.12 shows customer satisfaction initiatives

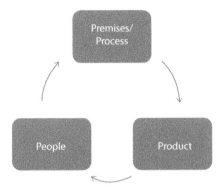

Figure 11.12 Customer satisfaction initiatives.

2. Product
- *Pre-emptive prevention, expansion prevention and mis-delivery prevention*: Boosting product-service quality.

3. People
- *Developing a comprehensive dealership training system*: Strengthening training programs to improve human resources and skill levels.

11.1.6   Vision for the Future

1. Future Initiatives
Structural changes surface in the economy because of awareness about environmental issues globally and the growth of developing countries have a significant effect on their business activities. With the growth of emerging economies, competition in the market has intensified, and online information is exerting a significant influence on performances. In future, this will require them to provide tailor made products to every region of the world more affordably and speedily.

2. Triple Zero
Zero CO_2 emissions will be guaranteed by using original renewable energy. Also, zero energy risk and zero waste will also be ensured with the collaboration of local communities. Figure 11.13 depicts Triple Zero and its coexistence with local communities.

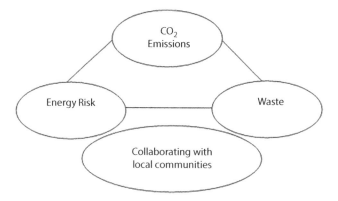

Figure 11.13 Triple Zero + coexistence with local communities.

11.2   Case Study in a Four-Wheeler Manufacturing Unit

This unit was established in Feb 1981 with the objectives of modernizing the Indian automobile industry, producing fuel efficient vehicles and producing indigenous utility cars for Indian population. It is the leader in the car sector, both in terms of revenue earned and volume of vehicles sold. Production of cars commenced in 1983. By 2004, over five million vehicles have been produced. Its manufacturing facilities are available at two locations that is, Manesar and Gurgaon.

11.2.1   Company Principles

It has adopted the norm of same fabric and color uniform for all its employees thereby, giving them an identity. In order to have no time loss in between shifts, employees reported early for shifts. It has an open office system and practices kaizen activities, job-rotation, teamwork, quality circles and on-the-job training.

11.2.2   Company Objectives

There was a need to provide a reliable, better quality and cost effective car to the customers. This was established in such a scenario with a resolve to bring about technological modernization and expansion of the automobile sector. This has been entrusted with the task of achieving the following policy objectives:

- Modernization of Indian automobile industry.
- For economic growth, large volume of vehicles had to be produced.
- For conservation of scarce resources, fuel-efficient vehicles were the need of the hour.

11.2.3　Company Strategy and Business Initiatives

For three decades, it has been the world's leader in mini and compact cars. Its technical superiority lies in its capability to pack performance and power into a lightweight and compact engine that is fuel-efficient and clean. This organization is clearly 'employer of choice' for young managers and automotive engineers across the country.

11.2.4　Technology Initiatives

Indian customer has passion for fuel efficiency when they have to choose automobiles. Achieving more energy for car from single drop of fuel is a challenge for the designer but is important for the economy, the planet and the customer. At the same time, a speed conscious, young and fast growing India demands better pick-up and instant response during acceleration. A third requirement is space efficiency, so that the car can cope with parking lots and congestion on roads.

The company's new K-series engines deliver on all these fronts. The organization believes that the technology's purpose is to serve mankind with products that use minimum resources and reach out to maximum customers, good for their long term safety, happiness, well-being, health and meet the needs of society. Better technologies, better thinking, better processes, more ideas and sensitivity that makes a difference in customer's life, help them develop better cars and thus, a better living. Today, the R and D team, has many achievements:

- It has launched many new models in India in the last few years.
- In India, some of the most fuel efficient petrol cars come with its badge.
- Launch of factory-fitted CNG variants. The factory fitted CNG (Compressed Natural Gas) vehicles use advanced i-GPI (Intelligent Gas Port Injection) technology. State-of-the-art i-GPI technology is used here.

- The new concept Single Minute Exchange of Dies (SMED) is being adopted. This helps in changing of die setup within single digit minute, thus, improving operating efficiency and machine utilization.
- Almost all its cars obey ELV norms, which means they can be fully recycled and are free from any hazardous material.
- Plastic Intake in K-series is an example of technologies adopted for light weight construction. Light Piston, Nut-less ConRod and Optimized Cylinder Block for light weight configuration, High Pressure Semi-return Fuel System, Smart Distributor Less Ignition (SDLI) with committed advanced injectors and plug top coils for better performance.
- Wagon R Green: Wagon R is a balance of performance, space and comfort in a new design. Its new model known as Wagon R Green is available on CNG. It ensures fuel-efficiency, safety, reliability and more power. CNG technology is another step for keeping low cost of ownership for customers.
- ESP (Electronic Stability Program): An onboard microcomputer displays vehicle's stability and behavior with sensors on a real time basis. During instability due to lane change or high speed cornering, it automatically applies differential brakes at the four wheels to keep the vehicle stable and on intended track without any additional driver involvement.
- Sequential injection has been introduced in LPG (Liquefied Petroleum Gas) vehicles to ensure reduced emissions, better fuel economy and improved performance.
- Variable Geometry Turbocharger (VGT) has been introduced in diesel engines for improving performance and fuel efficiency.

11.2.5   Management Initiatives

The organization develops a culture in which higher standards of individual's accountability, transparent disclosure and ethical behavior are ingrained in all business dealings and are shared by management, employees and board of directors. The firm has established procedures and systems to ensure that its board of directors are well-equipped and well-informed to fulfil their overall responsibilities and provides the management with strategic direction needed for creating long term shareholder value.

To meet the organizational responsibilities of safe working environment, the company has established an OHSMS (Occupational Health and Safety Management System) for:

- Managing risks – They identify all hazards by undertaking assessments, external and internal audits as well as all the necessary actions to prevent and control injury, loss, damage or ill-health.
- Complying with legal and other obligations – They ensure that business is managed in accordance with occupational health standards and safety legislations.
- Establishing targets and review mechanism – They manage their commitments by using coordinated safety plans and occupational health for each site and area. They tend to measure progress, leadership support and ensure continual

		MATERIALITY MATRIX		
SIGNIFICANCE TO THE STAKEHOLDERS	HIGH		Non-discrimination & human rights Road safety Skill development	Product safety Business growth & profitability People development & motivation Employee wages & benefits Occupational health & safety Customer satisfaction Process emissions Industrial relations Corporate governance Product quality Product emissions Compliance
	MEDIUM		Child and forced labour Green supply chain Green service workshop Information security & data privacy Product labeling	Atrition Water conservation Material optimisation Effluent waste Waste management Government policy and regulations Foreign exchange fluctuations
	LOW	Biodiversity Indigenous rights	Green products	Competition R & D capabilty Energy conservation Business ethics Macro economic factors
		LOW	MEDIUM	HIGH
		SIGNIFICANCE TO THE COMPANY		

Figure 11.14 Materiality matrix.

improvement. Health and safety performance is always among parameters for evaluation.

- Providing appropriate training and information – They provide all the necessary tools to all its vendors, employees, visitors and contractors to ensure safe performance at work.
- Ensuring meaningful and effective consultation – They involve all interested people and employees in the issues that harm health and safety at workplace.
- Communicating – They believe in the transparent relations of the performance and OHS commitments.
- Promoting a culture of safety – They believe that all incidents and injuries can be prevented and everyone is responsible for their own and their processes safety. All responsibilities are clearly defined for all personnel, managers and supervisors. Figure 11.14 shows materiality matrix.

11.2.6 Quality

The company is awarded ISO 27001 certification by Standardisation, Testing and Quality Certificate (STQC), Government of India. The quality management is certified against ISO 9001:2008 standard. These systems are re-assessed at regular intervals by a third party. To increase customer satisfaction through improvement of services and products, PDCA functions and levels of organizations are followed. Table 11.2 depicts the quality tools employed for realizing overall organizational objectives.

Table 11.2 Quality tools.

5S	4M	3M	3G
SEIRI — Proper Selection SEITION — Arrangement SEISO — Cleaning SEIKETSO — Cleanliness SHITSUKE— Discipline	MAN MACHINE MATERIAL METHODS	MURI - Inconvenience MUDA - Wastage MURA - Inconsistency	GENCHI — Go to actual place GENBUTSU — See the actual thing GENJITS — Take appropriate action

It recently enhanced its product range with an aim to meet customer needs. A striking new look and a more daring approach to design began with the launch of Swift Dzire and Swift. Another major development is its entry into used car market, where customers are allowed to bring their vehicles to 'True Value' outlet, where it can be exchanged it for a new one, by paying the difference.

11.2.7   Sustainable Development Goals

The **Sustainable Development Goals (SDGs)** or **Global Goals** are a collection of 17 interlinked global goals designed to be a "blueprint to achieve a better and more sustainable future for all". **Five** critical dimensions: people, prosperity, planet, partnership and peace, also known as the **5P's**. However, it actually refers to **four** distinct areas: human, social, economic and environmental – known as the **four** pillars of **sustainability**. Human **sustainability** aims to maintain and improve the human capital in society. The SDGs aim to "ensure that all human beings can enjoy prosperous and fulfilling lives and that economic, social, and technological progress occurs in harmony with nature." Inequality is one of the defining issues of this generation and requires a commensurate **focus** that, to date, has been lacking. Various SDGs are:

18. No poverty
19. Zero hunger
20. Good health and well-being
21. Quality education
22. Gender equality
23. Clean water and sanitization
24. Affordable and clean energy
25. Decent work and economic growth
26. Industry, innovation and infrastructure
27. Reduced inequalities
28. Sustainable cities and communities
29. Responsible production and consumption
30. Climate action
31. Life below water
32. Life on land
33. Peace, justice and strong institution
34. Partnerships for the goals

These **17 objectives are interrelated** and often the key to one's success will involve the issues most frequently linked to another. They can be summarized as follows:

- Eradicate poverty and hunger, guaranteeing a healthy life
- Universalize access to basic services such as water, sanitation and sustainable energy
- Support the generation of development opportunities through inclusive education and decent work
- Foster innovation and resilient infrastructure, creating communities and cities able to produce and consume sustainably
- Reduce inequality in the world, especially that concerning gender
- Care for the environment combating climate change and protecting the oceans and land ecosystems
- Promote collaboration between different social agents to create an environment of peace and sustainable development.

Develop sustainable, resilient and inclusive infrastructures; promote inclusive and sustainable industrialization; increase access to financial services and markets; upgrade all industries and infrastructures for sustainability; enhance research and upgrade industrial technologies. The remaining three targets are "means of achieving" targets: Facilitate sustainable infrastructure development for developing countries; support domestic technology development and industrial diversification; universal access to information and communications technology.

Implement the 10-Year Framework of Programs on Sustainable Consumption and Production Patterns; achieve the sustainable management and efficient use of natural resources; reducing by half the per capita global food waste at the retail and consumer levels; achieving the environmentally sound management of chemicals and all wastes throughout their life cycle; reducing waste generation through prevention, reduction, recycling and reuse; encourage companies to adopt sustainable practices; promote public procurement practices that are sustainable; and ensure that people everywhere have the relevant information and awareness for sustainable development. The three "means of achieving" targets are: support developing countries to strengthen their scientific and technological capacity; develop and implement tools to monitor sustainable development impacts; and remove market distortions, like fossil-fuel subsidies, that encourage wasteful consumption.

Some of the goals compete with each other. For example, seeking high levels of quantitative GDP growth can make it difficult to attain ecological, inequality reduction, and sustainability objectives. Similarly, increasing employment and wages can work against reducing the cost of living. On the other hand, nearly all stakeholders engaged in negotiations to develop the SDGs agreed that the high number of 17 goals was justified because the agenda they address is all-encompassing.

Environmental constraints and planetary boundaries are underrepresented within the SDGs. For instance, the paper "Making the Sustainable Development Goals Consistent with Sustainability" points out that the way the current SDGs are structured leads to a negative correlation between environmental sustainability and SDGs. This means, as the environmental sustainability side of the SDGs is underrepresented, the resource security for all, particularly for lower-income populations, is put at risk. This is not a criticism of the SDGs per se, but recognition that their environmental conditions are still weak. The SDGs have been criticized for their inability to protect biodiversity. They could unintentionally promote environmental destruction in the name of sustainable development.

The graph in Figure 11.15 shows the total income in the last five financial years. A gradual rise in total income has been witnessed during this phase. Total income is the sum of the money received, including income from services or employment, payments from pension plans, revenue from sales, or other sources.

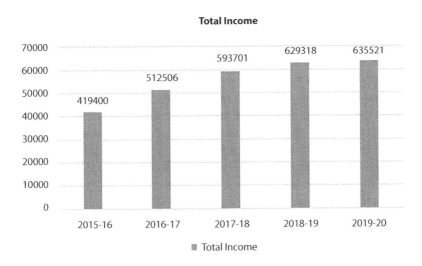

Figure 11.15 Total income for the last 5 years.

Several years after the launch of the SDGs, growing voices called for more emphasis on the need for technology and internet connectivity within the goals. In September 2020, the UN Broadband Commission for Sustainable Development called for digital connectivity to be established as a "foundational pillar" for achieving all the SDGs. In a document titled "Global Goal of Universal Connectivity Manifesto", the Broadband Commission said: "As we define the 'new normal' for our post-COVID world, leaving no one behind means leaving no one offline". Despite organizations adopting Sustainable Development Goals, Organization growth has improved over the years so leading to an improved profit

The sales and profit of the company have been shown in Figures 11.16 and 11.17 respectively whereas the growth of the company has been represented in Figure 11.18.

It has been analyzed through case studies that Sustainable green development and production techniques have an impact on overall performance of the organization. The factors **Sustainable Development, Green Manufacturing, Advanced Production Techniques (Computer Aided Manufacturing and Advanced Maintenance)** have a considerable impact as compared to others.

From the above data, it has been seen that the company sales have been growing. This is due to the introduction of new strategies and technologies in their products which is a prerequisite for survival in the market.

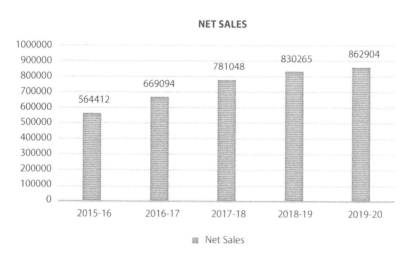

Figure 11.16 Sales for the last 5 years.

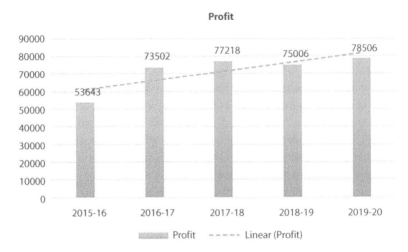

Figure 11.17 Profit for the last 5 years.

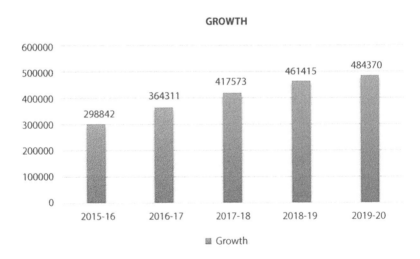

Figure 11.18 Growth for the last 5 years.

11.2.8   Future Plan of Action

- Continuously upgrading existing models.
- Developing products having alternative fuel options.
- Compliance with safety and emission regulations.
- Introducing new technologies.

- Developing knowledge of different automotive technologies through standard cost benchmarking and tables.

11.3   Conclusions

The research highlights the attributes of Sustainable Green Initiatives and Advanced Manufacturing and Maintenance Techniques and Performance of the Indian manufacturing organizations in order to enhance performance. The study also critically examines the Sustainable Green Initiatives and Advanced Manufacturing and Maintenance factors affecting the performance of these organizations. Moreover, the study illustrates the relationship of Sustainable Green Initiatives and Advanced Manufacturing and Maintenance Techniques and performance factors in Indian manufacturing industries to have overall business performance. The research critically examines the impact of Sustainable Green Initiatives and Advanced Manufacturing and Maintenance Techniques on strategy formation and thus organization's success. Finally, the research culminates with development of a Sustainable Green Initiatives and Advanced Manufacturing and Maintenance Techniques model for Indian manufacturing industry for sustained performance. The Conclusions from this study are:

1. Sustainable Development, Green Manufacturing, Computer Aided Manufacturing and Advanced Maintenance have emerged as significant contributors to sustained competitiveness in the organization.
2. Sustainable Development has a significant impact on an organization's competency as it involves innovation, sustained development, implementation and sustaining the performance.
3. Green Manufacturing plays a vital role in enhancing the Sustainable Green Initiatives and Advanced Manufacturing and Maintenance Techniques as it directly deals with manufacturing strategy planning, execution and controlling the operations. Its contribution to performance is higher than that of Sustainable Development.
4. Computer Aided Manufacturing directly affects the market value of a product and contributes significantly to the performance of an organization as it improves the productivity, quality and performance of the product.

5. Advanced Maintenance provides the much needed impetus for motivating employees towards organizational goals by making available the required resources which ultimately lead to the success and growth of any organization.
6. The qualitative sustainable green development model developed for Indian industry has the significant contributions of Sustainable Green Initiatives and Advanced Manufacturing and Maintenance Techniques factors to performance of an organization.

11.3.1   Limitations

The research limitations of the present work, presents suggestions for future studies. The samples in this study were collected among organizations. The validity of the findings regarding the relationship between Sustainable Green Initiatives and Advanced Manufacturing and Maintenance Techniques and organization performance may be hampered by as data on manufacturing practices and organizational performance were collected around the same point in time. At last, organization performance may be affected by other variables not accounted for in this work. It would be useful to examine the organization performance by taking factors like the legal situation and economic ones into account. Further limitations of the study are:

1. This study has been conducted in manufacturing industry only. Factors may vary according to manufacturing industries of different other products like bicycle, machine and machine tools, material handling equipment, Farm and agri-machinery.
2. The scope of this research has been limited to North India only; significance of issues may differ in other parts such as South, Central or whole of India.
3. A qualitative sustainable green development model has been developed. Other modeling techniques can be explored.

11.3.2   Suggestions for Future Work

The primary aim of this research is to synthesis performance concept and exploring Sustainable Green Initiatives and Advanced Manufacturing and Maintenance Techniques for manufacturing organizations, while a similar study can also be conducted in future for other Indian product, process and service industries as well. The work is aimed at developing Sustainable Green

Initiatives and Advanced Manufacturing and Maintenance Techniques and performance model for North Indian and auto parts manufacturing organizations and various manufacturers have been treated alike irrespective of the sector of manufacturing organizations. Another direction for future research is developing area-wise, sector-wise and product-wise model for manufacturing industry. Thus, individual case study could be conducted for different areas, products and sectors of manufacturing industry and accordingly the typical methodologies can also be evolved in future. While this research provided an insight into Sustainable Green Initiatives and Advanced Manufacturing and Maintenance Techniques and their relation with performance, it also discovered areas that could improve from further research. This work focused only on Sustainable Green Initiatives and Advanced Manufacturing and Maintenance Techniques of organizations. Future works could focus on other competencies as well. By doing so, a better and broader understanding of the effects that other competencies have on organization's performance may be accomplished.

1. Sustainable Green Initiatives and Advanced Manufacturing and Maintenance Techniques of manufacturing units have been explored. Future research could focus on the other competencies like innovation, design and development, supply chain, green manufacturing and such others.
2. Future research could also concentrate on sectors like large scale, medium scale, and small scale industries for model development in those industries.
3. Such studies can also be conducted in other regions of India to develop a holistic model for manufacturing industry.
4. A mathematical model could be attempted in future as in the present study a qualitative model only has been developed.

Index

Printed and bound by CPI Group (UK) Ltd, Croydon, CR0 4YY

27/10/2024

14580129-0002